# Remarkable Structures

Engineering Today's Innovative Buildings

# Remarkable Structures

Engineering Today's Innovative Buildings

**Sutherland Lyall**

**Princeton Architectural Press**
**New York**

**For Rosey**

Published in 2002 by
Princeton Architectural Press
37 East Seventh Street
New York, New York 10003

For a free catalog of books, call 1.800.722.6657.
Visit our web site at www.papress.com.

Text © 2002 Sutherland Lyall
This book was designed and produced by Laurence King Publishing Ltd
The moral right of the author has been asserted
05 04 03 02 5 4 3 2 1 First edition
Published simultaneously in Great Britain by Laurence King Publishing Ltd

Designed by Gavin Ambrose at Mono www.monosite.co.uk
Research coordinated by Jennifer Hudson
Edited by Lucy Trench

For Princeton Architectural Press:
Project coordinator: Mark Lamster
Project editor: Nicola Bednarek
Special thanks to: Nettie Aljian, Ann Alter, Amanda Atkins,
Janet Behning, Megan Carey, Penny Chu, Jan Cigliano, Jane Garvie,
Tom Hutten, Clare Jacobson, Nancy Eklund Later, Linda Lee, Anne
Nitschke, Lottchen Shivers, Jennifer Thompson, and Deb Wood of
Princeton Architectural Press—Kevin C. Lippert, publisher

Library of Congress Cataloging-in-Publication Data
Lyall, Sutherland.
   Remarkable structures : engineering today's innovative buildings /
Sutherland Lyall.
       p. cm.
Includes bibliographical references and index.
   ISBN 1-56898-330-1 (hardcover)
   1. Building. 2. Structural engineering. 3. Architectural firms. 4.
Buildings—Case studies. I. Title.
   TH845 .L95 2002
   690—dc21
                              2001006167

Printed in Hong Kong

# Contents

## Section 1

## Section 2

# Introduction

Some time towards the end of the twentieth century structural engineering shifted gear. An obvious part of the change was that structural engineering had made a significant move away from its traditional role of under-acknowledged handmaiden of the art of building. In the new alignment many structural engineers had become comfortable with the analysis and resolution of complicated, non-rectilinear building forms. For engineers that is more significant than it sounds. For in the 1950s, at the beginning of the change, the physics and mathematics of conventional engineering analysis still stood four-square on the foundations laid down by the great scientists of the seventeenth century Enlightenment: Galileo, Hooke, Newton and especially Leibniz, whose accessible differential calculus became the tool for the structural analysis of regular geometric building forms. Although Modernist architecture saw itself as pioneering new forms and materials, it was just as much an architecture of right angles and regular geometry as the architecture it rejected. So too, the new favoured Modernist materials had different moduli of elasticity and different bending, shear and deflection strengths from the conventional architectural building materials of the past; yet there was nothing particularly new about the way they were analyzed.

# New forms, new materials

During the latter half of the twentieth century, however, the canon of reputable architectural usage acquired non-orthogonal forms such as nets and grid shells and thin skins, warped surfaces, inflatables and geodesics. These sometimes employed new and unexpected structural materials: timber tailings, cardboard tubing, titanium, high performance fabrics, polymers and – a considerable surprise this – standard production-line glass. The structural types which resulted, often free form and complex, are most readily analyzed, and in some cases can only be practically analyzed, using computer applications developed during the same period.

It is true that the purpose of structural engineering is relatively prosaic: it is to satisfactorily resolve all the forces in a building and its support on the ground into a state of equilibrium. It is also true that most of the conceptually and mathematically challenging bits of structural engineering have answers which have been appropriated from associated engineering disciplines such as hydrology and aeronautics. One fabric analysis program, for example, is a creative leap from a program devised to model the topology of ground water. The program used to analyze the Bilbao Guggenheim structure was first used to design French Mirage aircraft and later Boeing 777s. Whatever the source, these and other programs, such as the home-brew program devised by the Czech checking engineers for the Prague Castle glasshouse, have to an extent released structural engineers from the grind of calculation and allowed them to think about buildings which would simply not have been possible in the pre-computer era. Happily in recent years architects' clients have been prepared, even eager, to accept many of these new forms and materials. In these circumstances the relationship between innovative architects and their structural engineers has become something more complicated than the architect handing over concept drawings for stability calculations by the engineer. Structural engineers have had in some way to become active participants in the creative process, or at least have had to develop an understanding of its nature, to enable that intense, collaborative relationship which seems necessary for the successful realization of avant garde architecture in the early twenty-first century.

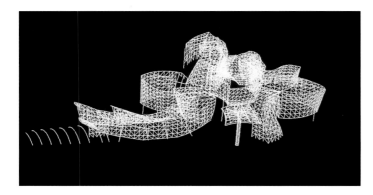

**Above:** Wire frame rendering of the Guggenheim Museum, Bilbao, by SOM and Frank Gehry, 1997.

# Mind sets

Inevitably what structural engineers do and the mind set they bring to the doing of it have undergone change. The great British engineer Ove Arup once made the proposition, much repeated with added irony by architect-hardened engineers in private, 'Why bring an architect down to earth when you can take him up to heaven?' This comforting, patronizing mantra encapsulates the old self-perceptions of the engineer as a kindly, all-knowing sorcerer, conservatively guiding the impetuous artist, the architect-dreamer, and bringing his fancies into concrete existence. This view lingers: for example, the engineers' report on the London Eye concept design regularly expressed mild astonishment that the architects had more or less got right all the basic structural issues.

But the Arup approach is no longer adequate for the relationship. Structural engineers have gained a new self-confidence, but in some cases it has become a prima donna arrogance exceeding leading architects' sense of their own importance. At the very least engineers have had to develop an empathy with the creative visual process and in some cases, such as that of the great contemporary European engineer-architect Santiago Calatrava, they have actually learned this in architecture school.

# The secret history

It needs to be said that the greater body of structural engineering analysis, like the greater body of architectural design, is to do with conventional, unexciting, unimaginative buildings. Bread and butter engineering calls for conservative, careful folk: ultimately the engineer's main professional task is to ensure that nothing much goes wrong. Yet, ironically, the history of engineering, or at least the interesting bits of its history, is a chronicle of risk-taking with structure and construction on an extraordinary scale. It has not always been very successful risk-taking either: much of the long-term history of engineering is a narrative of collapsed cathedrals, dams, tunnels, bridges. These many failures are of course redeemed by the fact that far more is learned from them than from structures which remain stable. So today's engineering may often call for heavy-duty mathematical computation and a high regard for safety, but it is ultimately an experimental activity whose failures continue to provide knowledge about both the art and the farthest limits of safe engineering.

## Devalued

Many of the structural engineers who have been involved with leading edge, celebrity building design have been conscious of the radical and sometimes extravagant strain running through the history of their art, also of the engineer's role in history. The late Ted Happold, the ebullient British engineering agitator, once wrote about how 'nearly everyone in Britain, and perhaps elsewhere too, thinks that creative design in building is exclusively due to architects.' He had just made the point that the

staggering development of technologies which had so changed the world in the previous 150 years had been started not by architects but by engineers. He also argued that the engineer's art was intensely creative 'in that it extends people's visions of what is possible and gives them insight'. The outcome was often a 'bare aesthetic' which was both new and related to natural rather than historical precedents. Here Happold staked a kind of claim to the moral high ground that Modernist architects had occupied for much of the century and which, during the time of Happold's writing, they had largely abandoned for the ironic neo-historicism of Post Modernism. The sparse, Calvinist, white wall and strip window aesthetic of International Modernism, so its architects had claimed, was the result of responding to the building's programme in terms of the structure and materials of construction. And this is what Happold seems to be appropriating to himself – or at least to structural engineering. It probably ever has been thus, for the internal principle that has guided engineers is doing much with little. As David Glover, Arup engineer for Future Systems' Lords media centre has put it very prosaically, 'As an engineer you don't like to see good materials wasted.'

Happold's view was not a lone one. Peter Rice, the engineering idol who died before his time in 1992, wrote, 'Unlike their predecessors, today's engineers work behind a screen of other egos... today's engineers are just as daring, just as inventive as their Victorian counterparts. It's just that even when we know about their work we don't have a label to attach to [it]... Engineers need identity. Engineers need to be known as individuals responsible for the artefacts they have designed.' The internal engineering literature has many articles on this theme and private conversation goes much further. But rarely are these views voiced in public.

## Delicate balance

The difficulty faced by engineers who take this view, even the prima donnas, is that the people who stand between them and the sun are architects, the people most frequently responsible for their appointment as consultants. So it is not entirely surprising that many engineers decline to discuss the issue or take the overly diplomatic line of describing how it works in reality rather than how it perhaps should go. Mark Whitby of the 360-strong British engineering practice Whitby & Bird, describes his understanding of the architect-engineer relationship thus: 'Structural engineers are like session musicians. They come into the studio to back the music of the best artists. And then hang around waiting for the next call.' So in the division of credit for any significant work of architecture the practical reality is that its engineer normally takes the backing musician's share. This is partly because the concept, the underlying vision of the building's form, is probably quite fairly assigned to the architect. Answering the question 'Who did what?' is never easy because architecture is increasingly the outcome of teamwork: the ideas of members of the group are taken up, passed around, developed by others and finally, both for commercial reasons and because nobody is really sure where to lay credit, are ascribed to the architectural partnership or a leading partner. It is very difficult, even within design teams, to really establish who did or thought of what. Small wonder that it is even more difficult to decide what the engineer contributed.

**Opposite:** The Millennium Bridge, London, by Ove Arup,
Foster & Partners and Anthony Caro, 2000.
**Above:** The Solférino Bridge, Paris, by Marc Mimram, 1999.

When the design achieves celebrity architects notoriously become greedy, effectively denying the engineer a creative role. When London's Millennium Bridge, an unusual horizontal suspension structure designed by the architect Norman Foster, the engineers Ove Arup & Partners and the sculptor Anthony Caro, turned out to oscillate alarmingly, the architect who had made all the running in the publicity about its opening suddenly drew back. He announced that it was an engineering issue, until press odium obliged him to shoulder his share of responsibility in public. But this works both ways. Engineers are not immune: the Arup engineer who had actually designed the bridge, Chris Wise, had left to set up his own practice and was thus never widely credited with the design (although he was billed as 'former Project Director' on the Arup website). He could therefore not, without public incredulity, be attacked for its failing. For a while the Arup website seemed to suggest aggrievedly that Marc Mimram, whose Solférino Bridge over the Seine had also oscillated, should have reported to the engineering world the causes of its wobble and his solutions: 'Those in the world that did know either did not make the events public or their chosen methods to inform others were limited and did not use the many direct avenues for informing the technical profession... There was no suggestion in any literature on the Solférino opening day that this could be a generic problem that could affect other bridges... The articles on the bridge were inconsistent and sometimes inaccurate with regard to the cause of the movement and the reasons for the closure of the bridge... No technical warning was issued by the designers of Solférino describing the phenomenon.'

A more frequent reality is that the architect-engineer relationship is simply not a particularly collaborative one and that from the beginning the engineer is denied the possibility of taking on a creative role. At the Bilbao Guggenheim Frank Gehry established the shapes of the elements and in a sense the task of the SOM engineers was merely to make it all work. Yet this required their creative and widely recognized engineering input, their lateral thinking and their recognition that it was all right to work on a building which lay so far outside the acceptable limits of stability and engineering normality.

## Beautiful and, if need be, useful

These issues might well never have emerged had it not been for changing fashions in the post-war period loosely centred around the kind of architecture called HiTech, its appearance determined by a high structural and mechanical engineering content. The early apotheosis of HiTech is the 1977 Pompidou Centre in Paris, designed by Richard Rogers and Renzo Piano with support from Happold and Rice, then both with Ove Arup & Partners. The celebrity structural engineering world is a small community.

Since then the HiTech style, with an intermission for Post Modernism, has segued into DeCon, Late Modernism and other more organic approaches to architectural form. But these later modes all pay a kind of homage to the memory of the machine aesthetic which was the foundation of the early twentieth century's revolt against the architectural historicism of the previous century. It is as if architects at the beginning of the new millennium are seeking to refine and renew their faith in what they believe to be the core of contemporary architecture via the clarifying fire of pure structure. It is also a reminder that some of the most daring designs of the heroic early period, such as Mies van der Rohe's frameless glass towers of the early 1920s and other dream designs of the early heroes can now, more or less, be built.

There is more to it than that, but current establishment architecture is characterized by an extensive reliance on structural engineering forms. In the process some architects have been known to urge their bemused consultants to produce more engineering-based structural solutions: 'archi-structure' as engineer Derek Sugden dryly labelled it. But even when this kind of design activity seems to involve art imitating an exaggerated form of technology, the structural engineer, if he or she is up to it, can probably correctly claim a major creative input into building design. Cecil Balmond, head of Ove Arup & Partners and evangelist for the notion of the structural engineer as creative designer, certainly takes this view. In an *Architecture d'aujourd'hui* interview of 2000 Balmond said, 'I thought we should intellectualize this [engineering] profession, bring it right up so it's not some kind of subservient field... I started challenging the architectural premise about space. I found the abstractions I offered gave insights into form and its organization. So we argue and debate. And with good architects this is not a problem. They value it.'

Many architects urge their structural consultants for more engineering-like solutions.
**Opposite:** Channel 4 headquarters, London, by Kenchington Little and Richard Rogers, 1994.
**Above:** Sainsbury's supermarket, Camden, London, by Ove Arup and Nicholas Grimshaw, 1988.

# Self-imposed limits

Even now engineers have not seen any particular need to include the history of engineering in their tertiary education syllabuses. It is not an entirely unreasonable position for there is still relatively little published on the subject of engineering innovation in the engineering literature, though there is a great deal about it in architectural texts. In the UK there is an engineering history group founded by a brilliant post-war engineer, the late Frank Newby, then head of F.J. Samuely & Partners, but its influence on the engineering community is difficult to assess. What is clear is that the resolute unwillingness of engineering schools to rate highly the teaching of engineering history means that they deny the profession a corporate memory and have saddled themselves with the absence of a tradition.

The world of architecture has consistently appropriated the most interesting historical engineers as honorary architects on the grounds of their creativity. In this scheme of thinking creative engineers rise above the dull number crunching of their fellows into the bright light of architectural enlightenment. It is a point of view which will prevail because, despite the eloquent protests of a few structural engineers, there seems to be no credible institutional alternative.

One of the traditional tests of professionalism is the existence of a written lore, a considerable body of writing in which the secrets and fundamental laws and mores and proprieties of the art are laid out for observation, discussion and self-criticism by the initiated. Architecture conforms to this model perfectly: it claims to have its beginnings at the dawn of civilization, with an early, sophisticated flourishing in the great cultures of the Near East, and a flowering in the empires of Greece and Rome. Its oldest extant literary text, Vitruvius's *Ten Books on Architecture*, is two thousand years old, written in the time of the Roman emperor Augustus. It is still read by (some) architecture students today. Ironically the contents of Vitruvius's book could almost as properly be described as engineering, though 'architecture' was the word Vitruvius used. In fact Vitruvius says that the three departments of architecture are making buildings, making timepieces and devising and constructing mechanical machinery, yet the last two activities have been resolutely overlooked by architects in any recent discussion of the *Ten Books*. Ever since, architects have been adept at changing the boundaries and definitions of architecture to ensure that whenever engineering or even mechanics impinges on it, that part of engineering becomes, *pro tem*, architecture.

Since the late twentieth century engineers have slowly begun to disseminate systematic knowledge among themselves of the creative genius of some of their predecessors. And they have begun to complain about their present lot. In the early 1960s the respected engineer A.J. Harris gave a talk to the Royal Institute of British Architects entitled 'Architectural misconceptions of engineering', in which he maintained that structural engineering was an art not a science, echoing the position enunciated in *The Philosophy of Structures* (1962) by the famous engineer Eduardo Torroja. Twenty years later Frank Newby curated an exhibition at the Architectural Association in London. Its theme, to use Newby's words, was the *art* of structures. Around a decade later Bill Addis's book *The Art of the Structural Engineer* appeared. Engineers were steadily plugging the engineering-as-art theme.

Harris's complaint was actually as much a complaint about misconceptions over the nature of scientific method. Scientists follow a series of rigorous procedures whose aim is to enable defining experiments to be repeated by other scientists. Central though it is in proving the credibility of scientific propositions, the dull repetition of experiments is not the way scientific *discoveries* are made. Like most other interesting, credible propositions such discoveries come about via a mixture of creative thought, sheer genius, serendipity, lateral thinking and accident. In short, the very things which lie behind creative engineering.

## Creative? Art?

But architects have slightly different meanings from engineers for 'creative' and for 'art'. For architects the engineer's creativeness is actually merely cleverness in analysis and synthesis, the engineer's art is mere artfulness. It is difficult, for example, to think of art always requiring dull repetitive calculation, as does even the best engineering flight of fancy. Stephen Morley, design principal with Modus and Sinclair Knight Merz, has an example: '...one can get the measure of a structural system quite quickly. The sizes of the main arches at Stadium Australia, spanning 286 metres [938 feet], were estimated in one line from the ubiquitous formula, WL/8E. It just took eight months to prove it.'

An example of engineering creativity is Buro Happold partner Ian Liddell's brilliant brainstorm in recognizing that the Millennium Dome's fabric structure – however perverse it might have been to create a compression form with a primarily tension structure – could be thought of and built in terms of straight lines. For architects this would not be classified as a creative act. For them 'creative' has almost always meant visually creative. More than a decade earlier architect James Gowan, in a book celebrating the contribution of engineers to the creative art of building, wrote, 'If a scrupulous distinction were made between architecture and engineering, it would be that one is concerned primarily with art, and the other, utility.' Derek Walker, then architecture professor at the Royal College of Art, had written in the same book, 'The sad reflection I have as an architect is that there are still too few engineers who can span the chasm that separates inventive design from humdrum mediocrity.' A decade and a half later something like this attitude remains among architects and the exceptions seem to prove the rule. Peter Rice was one of the exceptions. Richard Rogers, whose first major project with him was the design of the eminently engineering-like Pompidou Centre in Paris, wrote: 'Peter is not like any other engineer. He does not wait for the architect to develop his ideas and then offer options of how to prop them up... Like his great predecessors [he] is able to step outside the confines of his professional training, transferring technical problems into poetical solutions. His design combines order with delight, science with art.'

## Forebears

One of Walker's several exceptions to his generalized proposition about engineers was Frank Newby, then head of Samuely & Partners, who around the same time had curated the Architectural Association exhibition 'The Engineers'. It showed the work of a selection of the canonical nineteenth century structural engineers: Thomas Telford, Robert Stephenson, James Eades, Gustave Eiffel, the US skyscraper pioneer William Le Baron Jenney, the Swiss concrete bridge builders Robert Maillart and Eugène Freyssinet – and Newby's *lieber meister*, Felix Samuely. It was a carefully thought out show. But it was preaching to the converted. The theme and content might possibly have been startlingly novel to some engineers, but it was nothing particularly new to architects, brought up as they had been on a diet of these names plus, among others, those of the great English engineer Isambard Kingdom Brunel, the mid nineteenth century French engineer of iron and steel Henri Labrouste and the late nineteenth century concrete pioneers François Coignet, François Hennebique and Auguste and Gustave Perret. It says something about the importance of engineering for the early phases of modern architecture that quite as many engineers seem to figure in its pantheon of architectural heroes as do architects. And there is a reason for this, which lies in the way Modernist architectural history was written.

**Above:** The Tavanasa Bridge, Grisons, Switzerland, c.1905 (destroyed 1927), by the great bridge engineer Robert Maillart.

## The Giedion construct

An important section of the 1941 *Space, Time and Architecture* by the great apologist of Modernism, Sigfried Giedion, is a discussion about nineteenth century innovations in structural engineering at a time, according to Giedion, when architecture had become so trivialized that it was hardly worth discussing. He wrote, 'The seeds of the architecture of our [present] day were to be found in technical developments.' In addition, said Giedion, the nineteenth century threw up a number of new building types such as market halls, railway stations, department stores, the great exhibition buildings, long span bridges, towers. The new problems which they raised and their new solutions were not things with which conventional, history-bound architects were equipped to deal, nor would they have known what to do with the new materials, notably iron and steel and reinforced concrete and big sheets of glass. But engineers were exactly the people to deal with them. So the history of real architecture, Giedion implies, in that architecturally infertile century, was actually the chronicle of engineering invention.

Giedion was not about to lose his architectural audience by overstating the contribution of engineers. He quotes the Art Nouveau architect Henri van de Velde thus: 'The extraordinary beauty innate in the work of engineers has its basis in their unconsciousness of its artistic possibilities – much as the creators of the beauty of the cathedrals were unaware of the magnificence of their achievements.' This is not a description of the engineer as a thinking, creative being, rather as a kind of noble savage, the beauty of whose unselfconscious creations is accidental. Giedion makes a revealing aside, 'There are...those curving Swiss bridges which are formed out of thin slabs of ferro-concrete...which embody unexplored potentialities for architecture.' There is no hint here that these brilliant no-architect bridges are the work of the engineer Robert Maillart – although there is a long encomium on the engineer's work later in the book.

So the Giedion construct, followed in large by Nikolaus Pevsner and almost all establishment commentators and teachers of the early part of the twentieth century, ran something like this. Nineteenth century architecture had lost its way in a morass of revivalist styling; thus it and its architects forfeited any claim either to being taken seriously or to taking their place in representing the century's architecture. Engineering innovation had made all the running in the understanding of new materials, new building types and construction problems (prefabrication, steel, iron, glass, reinforced concrete and so on); so its history could be neatly slotted into the void left by the unworthy architects. It had to be remembered, though, that this was not to be viewed so much as a chronology of creative achievements as the opening of a great mine of brilliant engineering for discovery and extraction by the new architects of twentieth-century Modernism.

Giedion's writing was scarcely veiled propaganda for the inevitability of the new architecture and the intention of his argument was that, despite appearances, Modernism actually represented a clear historical continuity with the past. This was not the continuity of trivial appearance but a continuity established by the new tradition of structural (and to an extent mechanical) innovation. In the process the very visible differences between the architectures of the twentieth century and the past became the result not of wilful change for its own sake but of an inexorable interchange between new forms of construction and new structures.

## Fitness for purpose

Though patronizing, van de Velde's idea of the nineteenth century engineer as an unselfconscious child of nature innocently creating beauty by following the path of simple utility made a nice fit with this Modernist belief that beauty could (and, according to hardliners, automatically did) arise from following the sparse, functional processes of engineering, in which nothing unnecessary was introduced. As Giedion said of the great bridge engineer Maillart, 'In designing a bridge Maillart began by eliminating all that was non-functional; thus everything that remained was an immediate part of the structure.' Materials were used in accordance with their own natures and with the natural purpose of the structure in which they were used – all without decorating or distorting them. In the same way the beauty of plain forms and spaces arose from their perfect harmony with the activities they enclosed. It was as if, somehow, the technology created the aesthetic.

## The Machine

Equally engineering-orientated was the adoption by Modernist architects of the stance that the Machine – industrialization, mechanization, however it was described – was a central inspiration for form. More than that the machine was the emblem for the modern age of new materials and techniques (and social attitudes), it was a representation of rational thinking. The anticipated new forms and spaces were refined versions of the new building types of the nineteenth century, together with other, twentieth century structures: factories, power stations, airports, hospitals, schools, universities, mass housing and the great transport interchanges envisioned by the Italian Futurist, Antonio Sant'Elia. And the materials were the new materials produced by industrialization: as the Modernist British architect Wells Coates argued in the early 1930s, 'The most fundamental change in our technique is the replacement of natural materials by scientific ones.' It needs to be said that for all these references to scientificism and engineering what actually defined the appearance of most Modernist buildings was a standard palette of materials: white concrete walls, pilotis, flat roofs and metal strip windows. It was so de-rigueur a palette for Modernist architects that in 1932 the young architectural curator of the New York Museum of Modern Art, Philip Johnson, called his proselytizing exhibition 'The International Style'. Modernist architects were enraged at this less than heroic appellation for an architecture that claimed a special affinity with the deep ethics and mores of sternly rational engineering.

## A defining moment

For the early post-war world it was probably the Sydney Opera House which most re-established the eminence of the engineer in the process of serious building. Winner of the 1957 competition for the building, the Danish architect, Jørn Utzon, had drawn great, delicate concrete shells as the covering for the auditoria. At the time there was an assumption among architects that these would be built in something like the ferro-cement which had been pioneered by Pier Luigi Nervi in the 1930s and in the post-war years by the Mexican designer Felix Candela. Their designs, in cement reinforced by multiple layers of wire mesh, relied on shape for much of their stability – and a number of boats and leisure craft have been constructed from this material as well as buildings. So for most architectural observers the shapes of the opera house sails looked eminently buildable. What they forgot was that although the technique could be carried out at a small scale, the vast scale of the structure on Bennelong Point brought limiting factors into play that would have been prohibitively complex and expensive. Utzon suggested other possibilities but it required the intensive work of the Ove Arup & Partners' team to develop his final suggestion that the geometry of the sails could be thought of as parts of the surface of a sphere. The end result is plainly, almost shockingly, a heavy engineering solution. The probability is that a present day structural engineering solution would have embraced the fragile lightness of the original design. Nevertheless the Arup solution has remained enduringly impressive for the non-architectural world and the significance of the structural engineer's role in great architecture was firmly established.

**Above:** The distinctive concrete shells of Jørn Utzon's Sydney Opera House,
1957-73.

## A chronology

Cecil Balmond is author of a fascinating book, *Number 9: The Search for the Sigma Code* (1998). In exploring the ubiquity of the number nine in certain kinds of highly specialized mathematical series, it falls somewhere between the mystic numerology of the Cabala and those popular books of tricks and fun with numbers. As head of Ove Arup & Partners he has also been engineer of choice to such architectural luminaries as Zaha Hadid, Rem Koolhaas and Daniel Libeskind. In the *Architecture d'aujourd'hui* interview he argued that the three twentieth century architectural giants, Mies van der Rohe, Le Corbusier and Frank Lloyd Wright, had 'grabbed the [twentieth] century and, by 1960, the imprimatur on intellectual rights on design were wholly appropriated by the architectural establishment…It was a self sustaining thing…[nurtured by an unholy alliance between architects and their press]. But technology was changing. Through the 30s, 40s and 50s it was okay; with technology moving slowly everyone could catch up. But then came the acceleration of the 70s and then the computers taking over in the 80s. And suddenly light and air are also substances like structure, all in the control of technology…'

# Little masters

During the early decades identified by Balmond the structural engineering world was populated by a small cadre of Modernism-friendly engineers such as Pier Luigi Nervi of Italy, Ove Arup and Felix Samuely in England and Eduardo Torroja in Spain. In the 50 years after World War II many more talented individuals appeared on the scene. The engineering division of the great US architectural firm Skidmore Owings & Merrill developed its own luminaries, among them Myron Goldsmith and Fasler Khan. There are the US practices of Ty Lin, Le Messurier Consultants and Matthys Levy at Weidlinger Associates. There was Felix Candela of Mexico, who for a time in the 1950s and 1960s excelled in very thin concrete shells; the young, architect-friendly engineer Anthony Hunt, still active; Samuely's successor, Frank Newby; and such individual Arup engineers as Ted Happold (later to set up Buro Happold) and Peter Rice who founded the French multidisciplinary practice RFR before his untimely death. Frei Otto established a towering reputation for radical structures, one of the earliest and best known of which was the acrylic-clad cable net roof for the Munich Olympics in 1972. At the same time there was the work of such engineering academics as Mario Salvadori, the great teacher of structures and apostle of building systems concepts, the recently retired Arup mathematical guru Toshihiku Kimura and Peter McLeary, whose developments in structural theory and analytic methods have begun to support what has hitherto been a largely empirical science. These were the engineers who were most notably able to deploy an architect-sympathetic mode of thought. In doing so, they could absorb all the new radical thinking about engineering and, whether as partners in design teams or in personal interaction, disseminate it around their architect network.

Looming behind all these was the charismatic presence of Buckminster Fuller who even in the 1930s, when he began his public career as an inventor, was not so much an engineer as a prophet of avant garde structure. Deploying animated lectures, a series of patents for structural systems and shock aphorisms (such as 'How much does your building weigh?'), he harried the increasingly moribund post-war Modernist architectural establishment into thinking more interestingly about buildings and architecture – specially lightweight, possibly mobile structures. Widely credited with the invention of geodesic and tensegrity structures, in fact he creatively adapted the original ideas and patents of others – and for that reason, apart from his scant formal engineering qualifications, has always been held in a degree of suspicion by the engineering world. Yet far more than any engineer had managed to do, he impressed on whole generations of young post-war architects the possibility that buildings could be lightweight structures whose character was light and tension.

**Below:** Buckminster Fuller's 'Climatron' geodesic dome at the St Louis Botanical Gardens, 1960.

# New kinds of structure

Fuller's geodesics and the ultra-thin concrete shells of Nervi, Torroja, Ove Arup, Nicholas Esquillan and Felix Candela were all based on regular geometry so the curving surfaces could be analyzed in terms of familiar mathematical formulae. An exception to this is the work of the Swiss engineer Heinz Eisler. His thin shells are free form and their analysis has had to be based on model studies, interestingly often carried out by hanging model structures upside down and allowing them to freeze overnight. He has used the same technique to create large temporary ice structures on light fabric formers – or in the case of giant ice bubbles, very large balloons which are deflated once their surface has been sprayed and the water frozen. Although Eisler continues to build thin concrete shells they had fallen from popularity by the late 1970s because of the complications and expense of building formwork. Other factors in their demise were the emergence of membrane structures in the hands of Frei Otto and Ted Happold and the fabric structures devised by young designers such as Happold disciple Ian Liddell, the German-American Horst Berger and former partner David Geiger, with later followers such as Todd Dolland of Future Tents. Berger has the world's largest lightweight roof structure to his credit, the 1981 Haj Terminal at Saudi Arabia's Jeddah International Airport. In his *Light Structures, Structures of Light: The Art and Engineering of Tensile Architecture* (1997), he argues that architecture is the raising of technology to an art form to create the spaces which house human activity. 'Tensile architecture achieves this with one integrated structural surface, accomplishing all the things which, in conventional buildings, require the combination of many additive elements.'

## Membranes

Membrane structures could have quickly become an engineering-architectural cliché were it not for the almost infinite variety of ways in which it is possible to cover large areas so inexpensively, with minimal fabric materials, steel masts and, normally, concrete anchors. The main cost and the main difficulty lie in the interface between the edges of the fabric and the ground, whether that is a perimeter wall or some kind of fabric air seal. It needs to be said that the design of tensile membrane structures is almost entirely dependent on the use of computers because they are normally non-linear and thus defy classical analysis. The first use of digital computers in designing a membrane structure is claimed by David Geiger Associates in developing the air-supported roof of the US pavilion at Expo '70 in Osaka. Because there is a variety of approaches to form finding and analysis, computer programs tend to be developed by university research departments or by the big specialist engineering practices, from the Happold-originated TS Form, through Geiger Engineers' own Fortran-based software. There are, naturally, not many such programs in existence, perhaps ten or a dozen, although Geiger people claim that none really covers the whole conspectus from form finding through analysis to pattern cutting.

For a time in the 1960s it seemed that inflatable structures had a major role to play, but because of a well documented history of accidental deflations and the perceived need for constant supervision of air blowers they have not turned out to be as popular as it was hoped. Yet there have been some notable structures in the form of the landmark indoor tennis courts on New York's East River and the enormous Tokyo Dome of 1988, created from a series of thick inflated ribs and designed by Nikken Sekkei.

## Glass

As Cecil Balmond's chronology correctly suggests, by the end of the 1970s the engineers had loosened up and begun to investigate a whole array of new structural types. These ranged from grid shells, through cable nets, suspension structures, pillows and at least one semi-monocoque, to various hybrids in which two separate structural systems are used in conjunction with each other – such as Jörg Schlaich's glass vault over Hamburg's historical museum in which a glazed grid is given lateral resistance by a separate cable net slung diagonally underneath. It was as if structural engineers had thrown off a restraining cloak and commenced a frenzy of investigation into structures which five years before would have seemed either implausible or beyond the proper interest of engineers. What is significant is that these studies of new structure were, often enough, conceived in a spirit of independent radical inquiry rather than as a result of giving reality to visual solutions dreamed up by architects. It is not that architects were no longer important, simply that engineers had begun to develop a repertoire of structural forms which were their own – and which, with the increasing sophistication of both computing analysis and form finding, were beyond the ken of architects.

Until the 1990s the big glass manufacturers had concentrated on such things as decreasing the thermal emissivity of glass, fritting with hard-to-see ceramic dots in the name of solar shading and developing methods of supporting glass in big walls without traditional transoms and mullions. Glass had a special place in the early Modern Movement. It was associated with health and efficiency. The early kindergarten (children-garden) theorists made a quite literal connection between glasshouses, in which young plants grew under the life-giving sun protected from the elements, and the growth and nurturing of children. And there were two enduring Modernist icons of transparent architecture. One was Paxton's 1851 Crystal Palace, literally a cathedral of glass – though it was all supported by iron frames. The other was Mies van der Rohe's 1921 design for a glass tower. This was a multi-storeyed building comprising floors whose external skin was an ethereal veil of glass, which unlike Paxton's iron framed structure, was without visible support. From the 1970s it was possible to use unframed glass panels sealed with clear silicone as very big walls. Given the architectural history and the intriguing possibilities of constructing in glass, it is not entirely surprising that engineers and architects sought to test the material to its limits. British glass exponent Tim McFarlane says he once happened to look at bending moment and shear force performance figures from a glass manufacturer and was astonished to find them not all that different from those for conventional materials. In fact ordinary glass is very strong – but is prone to brittle fracture due to even minute surface damage. Even laminated float glass can crack around holes. But by using toughened glass which is bonded and laminated in various ways it was possible by the mid 1990s to think of an all-glass architecture.

**Opposite:** The lightweight tent structures of the Haj Terminal at Jeddah International Airport, Saudi Arabia, by Horst Berger, 1982.
**Above:** The glass tube Corporation Street footbridge, Manchester, by Ove Arup and Hodder Associates, 1998. Explorations into glass technology meant that by the mid-1990s all-glass architecture was a practicable option.

## Brick

Even brick structure was rethought for specialist large scale applications by the little known Uruguayan engineer Eladio Dieste. His preoccupation was with free-standing reinforced brick barrel vaults, with single curvature forms which are either semicircular or catenary in cross section and have no end bracing. Unlike traditional masonry barrel vaults these are, in Dieste's case, seated on restraining concrete frames at eaves level rather than on massive masonry supports and the result has been marvellous warped-plane roofs and, later, walls, in reinforced brickwork. The vaults produced by Dieste were constructed using timber formwork, much in the way a concrete shell would have been formed. But the only cement to be seen was in the form of traditional mortar laid between the bricks, which actually shrouds the steel reinforcement rods.

## Timber

Timber has been a construction material for so long that it seems hardly susceptible to radicalization. But for some years Buro Happold has been working with a number of different architects at Hook Park in the west of England creating buildings whose structure is made from timber tailings – effectively the waste left over after coppicing woodlands. More recently the centre, once described as 'an oasis of one-off exercises in timber gymnastics', has become more conservative but Buro Happold has moved in the direction of grid mats created from flexible, recently-cut green oak. In mainland Europe Julius Natterer has specialized in timber construction, including double curvature ribbed shells made from bolted and glued and often nailed laminated timber.

**Above:** The timber ceiling at Westminster Lodge, Hook Park, Dorset, by Buro Happold and Edward Cullinan, 1996.

**Opposite left:** A model of Santiago Calatrava's Swissbau Concret pavilion, 1989.
**Opposite right:** The Volantin footbridge, Bilbao, by Santiago Calatrava, 1997.

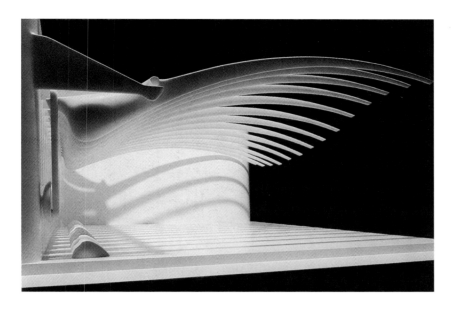

# The difficult case of Santiago Calatrava

In all this rewriting of the relationship between architects and engineers there is the difficult case of the engineer-architect Santiago Calatrava. A Valencian, he trained as an architect in Franco's Spain and then read engineering in Zurich, where he set up in practice in 1981. His early fame rests on the Stadelhofen railway station in Zurich of 1990, some fascinating mobile mechanical concrete constructions such as the Swissbau Concret Pavilion of 1989 and the Kuwait Pavilion of 1992 – plus a series of increasingly spectacular bridges which effectively re-wrote the lexicon of bridge forms. His Lyons railway station, a hybrid concrete and steel structural tour de force, established his credibility as an architect - though not many architects seemed to much like it. Since then he has designed and built an astonishing number of bridges and very big concrete buildings whose wilful skeletal forms have dismayed both architects and engineers – and architectural commentators who, preferring, naturally, to be able to deploy currently fashionable categorizations, are unable to pigeonhole his work. Is it perhaps Post Expressionist? It has some vague affinities with the small oeuvre of the architectural Expressionists of the early twentieth century – of which Mendelsohn's expressive Einstein Tower at Potsdam was one of the very few built examples. In a 1997 issue of *Werk Bauen + Wohnen* an editorial agonized over the problem and spoke of 'the quintessence of loading and load-bearing in expressively tortuous images.' Calatrava seems happily oblivious to these problems. His argument is that the languages of structure and geometry are of similar importance in his thinking about design – together with the nature of materials and of nature.

Of his favourite material, concrete, he has said that although it is universally available it is a 'difficult material and requires a great deal of expertise. And by this I mean not just technical knowledge, but also an understanding of the inner potential for poetic expression that materials possess.' For the early Modernists concrete was the prime material of the architectural tabula rasa, poetic only in that dry minimalist sense that it represented a kind of industrial purity.

And of his skeletal and organic shapes Calatrava talks of nature as his guide: 'There are many lessons one can draw from nature, real guiding rules and metaphors from observing plants and animals. To me, there are two overriding principles to be found in nature which are most appropriate for building: one is the optimal use of material; the other is the capacity of organisms to change shape, to grow, and to move.

'I have built tree-like structures and frequently my designs recall the form of skeletons. Behind this is the principle of recurrence. Whether in the case of trees or vertebrae, one finds the form dictated by the universal structural law that the base is thicker than the crown. The recurrence of this principle expresses economic efficiency. But it also arises from something beautiful, namely rhythm – the rhythm one finds in musical compositions.'

Small wonder then that engineers, particularly his outspoken critic, engineer Michel Virloguex, designer of the Pont de Normandie and an apostle of pure rationalism in bridge design, have as much difficulty with Calatrava's wild plastic imagination as do architects. The grudging, tight lipped half-acceptance of his work by both architectural and engineering fraternities has more than a whiff of the Inquisition pondering heretics who believe in too many things at the same time.

Two London examples of the current participation by architects in bridge design.
**Above:** Royal Victoria Dock Bridge, by Techniker and Lifshutz Davidson, 1998.
**Below:** West India Quay pontoon bridge by Anthony Hunt and Future Systems, 1996.

**Above:** The minimal Suransuns footbridge, Viamala, Switzerland, by Conzett Bronzini Gartmann, 1997.

## Bridges

Not all sensational structural engineering has much to do with architecture, though a number of recent bridges have been the work of architects in close collaboration with engineers. The Gateshead Millennium Bridge across the Tyne at Gateshead, near Newcastle in north-east England, was designed by a team of engineers, Gifford & Partners, and the architects Wilkinson Eyre. The first of its kind in the world, it is made up of two elliptical arches spanning the river attached to the same pivot on each bank. One arch forms the footway supported by cables from the other which is aligned at right angles to the footway. When boats need to pass underneath, the two pivot together until the supporting cables are parallel with the river, enabling a 25 metre (82 foot) high clearance under a 35 metre (115 foot) wide clearance zone in the middle of the arches.

The active participation by architects in bridge design, particularly in Britain, was a development of the end of the twentieth century – at a time when there seems to have been a worldwide demand for new bridges. But it is not invariable and some British engineers, at least, feel they are now in a position to operate as independent creative designers in their own right. In this they are working from structural first principles which, runs the old Modernist belief, are a sure recipe for beauty. This is certainly the case in the utterly minimal Suransuns footbridge by Conzett Bronzini Gartmann.

**Above, right and overleaf:** The pivoting Gateshead Millennium Bridge, by Gifford & Partners and Wilkinson Eyre, 2000.

## A tentative conclusion

The active participation of engineering in the architecture of the first part of the twenty-first century seems secure. That is partly because the number of structural engineers who are ready to participate in the rituals (and occasional self-delusions) of creative architectural design is measurably larger than half a century ago. It is also because the current fashions in architecture incorporate a high level of expressive engineering structure. Hence the relationship is contingent on styles of architecture not veering too far away from that model. But it may be that structural engineering's re-found capacity for radical thought and its new tendency to seek out solutions to problems which it, rather than architecture, has raised, may find creative expression (in the architectural sense of creative) in new structures such as bridges.

On the other hand, as professionals in the construction industry increasingly seek new alignments and relationships, it may be that RFR, the Paris practice founded by the late Peter Rice which incorporates the whole range of design skills, will be a future model. Another model is that of the movie or television production unit run by a producer with a creative director which lasts as long as the project and disbands or reforms elsewhere. Whatever models emerge, it seems that engineering needs to start investigating its own long history – and celebrating it.

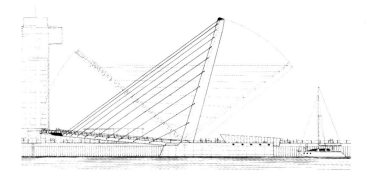

# Case Studies

Park keeper's flat and public lavatories, Shungu-cho, Hyogo Prefecture

Millennium Dome, Greenwich, London

Japan Pavilion, Expo 2000, Hanover

Eden Project, Bodelva, Cornwall

Great Glasshouse, National Botanic Garden of Wales, Carmarthenshire

William Hutton Younger Dynamic Earth Centre, Edinburgh

Rose Center for Earth and Space, American Museum of Natural History, New York

Tokyo International Forum, Tokyo

Mercedes Benz Design Centre, Sindelfingen

Lords Media Centre, London

Spandau Station, Berlin

Tower of Babel, Millennium Dome, Greenwich, London

Pavilion, Broadfield House Glass Museum, Kingswinford, West Midlands

Orangery, Prague Castle, Prague

City of Science, Valencia

Waterland, Burgh-Haamstede

Guggenheim Museum, Bilbao

Carlos Moseley Music Pavilion, New York

London Eye, London

Stock Exchange and Chamber of Commerce, Berlin

Educatorium, Utrecht University, Utrecht

Traversina and Suransuns footbridges, Viamala Gorge

Solférino Bridge, Paris

Expo Roof, Hanover

Hall F, Charles de Gaulle International Airport Terminal 2, Paris

# Engineer: **TIS & Partners**

## Park keeper's flat and public lavatories, Shungu-cho, Hyogo Prefecture, 1998

### Architect: **Shuhei Endo Architect Institute**

**The park keeper's flat and public lavatories are located in a small park adjacent to newly built elementary and secondary schools in the mountains of Hyogo Prefecture, one hour by bullet train from Osaka. The facility comprises a janitor's flat and male and female public lavatories, wrapped around by a set of galvanized corrugated iron spirals. The architect, Shuhei Endo, modestly describes this as 'Halftecture' (half+architecture).**

It is, he says, 'characterized simultaneously by both openness and closedness... Interior walls double as exterior ceilings and floors, which also extend as exterior walls and roofs and once again turn into interior parts. The interior and exterior form a linkage of changes, challenging architectural norms expected by the observer, and suggesting a new, heterogeneous architectural form. The facility is also a small attempt towards a new architecture realized by continuous interplay between the interior and the exterior and the interactive effect of partial sharing of roofs, floors and walls.'

  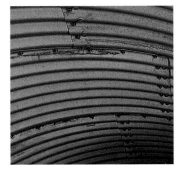 

Corrugated iron has hitherto been the special preserve of the Australian architect Glen Murcutt, who uses this universal Antipodean roofing material in new forms. Endo takes the material and its properties much further, in the sense that the corrugated iron is both wall and roof and floor and the accommodation exists as a separate entity within the spaces created by the metal.

The spiral is supported in a variety of ways, notably by a series of beams held on steel tubes, one pair for each element in the spiral. Occasionally the shape is maintained with arrays of seven or eight horizontal struts, while the base of one spiral is supported by a cantilever beam. The major engineering problem, which took a long time to resolve, was to work out how transient wind loads would affect the curving surfaces. In the end, the resolution relied on the fact that the sheeting was a secondary material and not part of the primary structure.

Corrugated iron comes in sheets small enough to be handled by one operative. It is a flat, galvanized steel sheet which is post-formed by an array of rollers. The corrugations give the originally floppy sheet great rigidity in the direction of the corrugations. In the mechanical process of creating the corrugations it is very simple to give the sheets a precise longitudinal curvature. The three dimensional twisting of Endo's spiral structure is achieved by cutting the relatively small sheets on the bias, then bolting adjacent pieces together in a way which allows each one to slip along the other, thus creating the twist. This is a completely standard method of manufacture and the structural engineering involved is essentially the relatively simple task of designing the props and beams for earthquake loads. Its most complex issue is the geometry of the spirals (which are sections of the circumference of a group of circles with differing diameters), and thus the individual sheets and their location in the overall form during construction.

Park keeper's flat and public lavatories

**Page 31:** One of the most prevalent of building materials used in a transforming, structural way.

**Opposite:** Like a ribbon peeling from a giant pencil sharpener, the TIS/Endo corrugated iron sprawls along the site.

**Above:** The functioning parts of the structure are inserted almost arbitrarily into the sweep of the sheeting. This is not double curvature structure so there is need for a post and beam support system, while the horizontal props on the curve to the rear right help maintain the correct curvature.

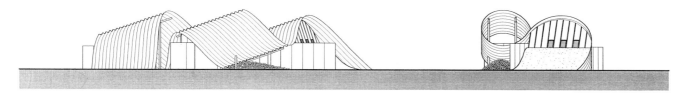

**East elevation**

**North elevation**

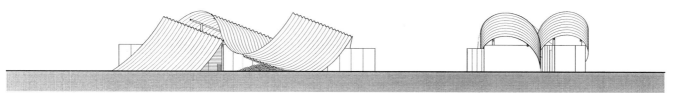

**West elevation**

**South elevation**

Park keeper's flat and public lavatories

This is a building on the far visual extreme and yet it involved very little structural design input. There is, for example, no special significance in the use of corrugated sheet steel since the corrugations serve merely their traditional function of providing local stiffness. The supporting structure is perhaps odd but it is actually a simple metal version of the familiar post and beam. And the sheets are bolted together just as they are when used as an industrial roofing material. Before this Endo executed several similar structures, one a bicycle shed beside a busy commuter railway line. Its stability was more or less achieved by suspending weights at the end of the corrugated iron sheets. When trains pass the whole structure shivers pleasantly. Subsequent unbuilt designs have explored the possibilities of spiraling shells in thin concrete.

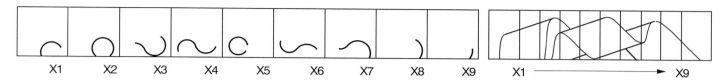

X1    X2    X3    X4    X5    X6    X7    X8    X9            X1 ———————→ X9

**Opposite:** What appears to be an unusually heavy support structure is a consequence of Japan's earthquake-related building codes. The pre-curved sheets of corrugated iron are simply bolted together side by side and end to end.

**Above:** A series of cross sections through the ribbon.

**Below:** Concept model which also tests the scale of the structure.

  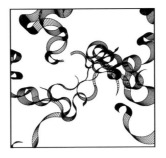 

**Above:** Studies in developing the geometric complexities of random ribbon arrays.

**Below:** Study to determine form using literal ribbons wound around cylindrical formers.

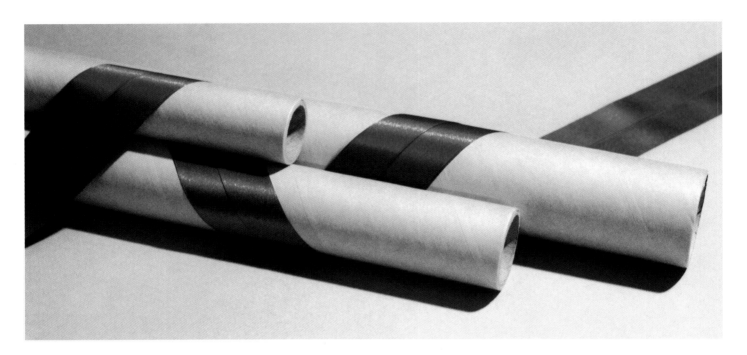

**Below:** Plan.

1 – Lounge

2 – Men's room

3 – Women's room

Park keeper's flat and public lavatories

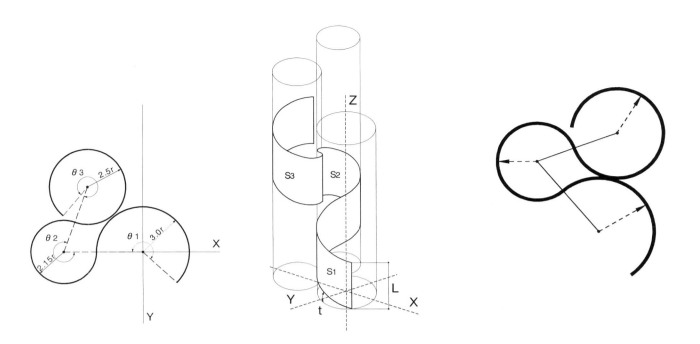

**Above:** Studies of the curving geometry of the ribbon.

$x = r \cos \varnothing$  $y = r \sin \varnothing$  $z = 2\pi r \varnothing n/360 \tan t + L$

**Below:** Park keeper's flat with earth berm and clerestory windows.

# Engineer: **Buro Happold**

Millennium Dome, Greenwich, London, 1999

## Architect: **Richard Rogers Partnership and Imagination Ltd**

**The events surrounding Britain's Millennium Dome turned into a major fiasco, with dramatically fewer visitors than projected and a series of profoundly embarrassing errors of management. Located on a heavily polluted peninsula near the Greenwich Observatory, the dome structure became synonymous with that failure. Yet as a structure and feat of engineering bravura, the dome was and has been a great success. And, unlike practically everything else to do with the project, it was completed on time and on budget. An enclosed space with a diameter of 320 metres (1049 feet) and an area of 80,000 square metres (860,000 square feet) it is the largest fabric structure in the world.**

In the late 1990s Richard Rogers' partner Mike Davies was in discussion with Gary Withers, creative head of the design practice Imagination, over a scheme – based on the theme of time – for housing the British Millennium exposition in a series of large pavilions. It was already running late and the whole show faced abandonment if a convincing solution was not in prospect. Davies rang Ian Liddell, a founding partner of Buro Happold, and asked him if he could design a 365 metre (1197 foot) fabric dome. Liddell, a pioneer in fabric design, was comfortable with the idea, though it meant that he would have to get a design worked out and the project out to tender in less than five months, with completion only 15 months later.

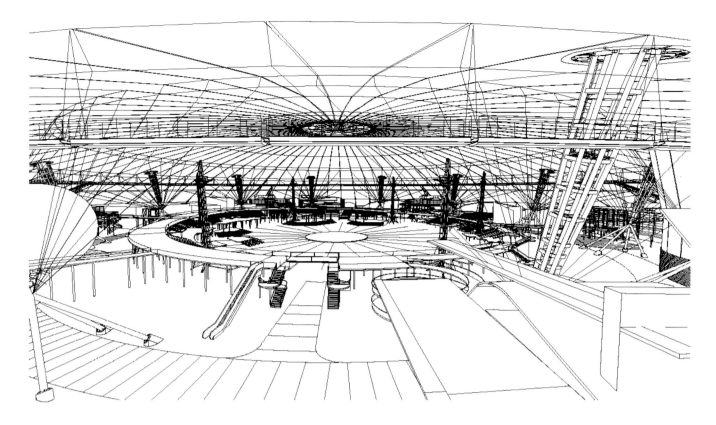

**Preceding page:** London's vast Millennium Dome, the entirely successful structure which housed the doomed British Millennium celebrations.

**Above:** Early perspective view of the dome interior.

**Below:** Plan of the dome. The Thames is on the right curving around the dome promontory to the left. The circular shape just below 9 o'clock is a ventilation tower for the Blackwall tunnel beneath.

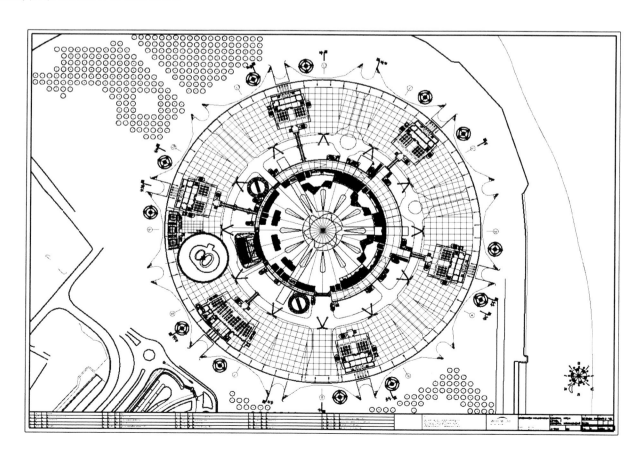

**Above left:** Part of the internal structure surrounding the central performance space with theme areas beyond.

**Below:** Section through dome.

**Above right:** One of the masts before final rigging. These enormous pylons were seated on open pyramids of tubular steel and are, here, stayed temporarily waiting for the fabric to be put up.

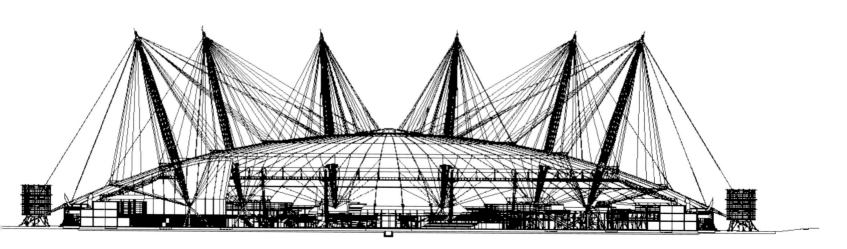

Domes are structural systems developed from the behaviour of quite small pieces of heavy material, traditionally masonry or brick, whose geometric arrangement allows them to span quite large spaces. The load generated by the weight of the material is normally the most important load and the stresses are resolved around the perimeter of the dome by massive encircling walls, buttresses and sometimes chains. Fabric structures can be of many shapes but the classic form established by the German architect-engineer Frei Otto is a minimal surface with equal tensions under pre-stress: to use Buro Happold's words, well curved surfaces with opposing curvature. So a dome is not a particularly 'natural' shape for a fabric structure, though its shape was appropriate to the theme of time and universality. It was, for a structure with overtly symbolic importance, also an easily understood shape, reminiscent of the Dome of Discovery, the centrepiece of the 1951 Festival of Britain.

One suspects that it was the fact that it was not a classic fabric design shape that intrigued the engineers. They had already designed and built several fabric structures based on the principles of a marquee tent, whose fabric is essentially flat and which rely on one or a number of masts.

The first sketch designs were for a 400 metre (1312 foot) cable structure with two rings of masts, a central ring of 12 and an outer ring of 24, with the forces transmitted, tent-like, to raking ground anchors. This soon developed into the current shape – a structure 320 metres (1050 feet) in diameter and 48 metres (157 feet) high, with a single ring of 12 masts located closer to the perimeter. The masts were now mounted on 10 metre (33 foot) high steel tripods, with an inner steel tension ring 30 metres (97 feet) in diameter to which the radial cables were anchored. The outer tie-down cables were replaced by external flying struts. Soon afterwards the height of the masts was increased, the flying struts were eliminated and a 1 kilometre (1094 yard) long ring beam introduced to take the horizontal component of the forces rather than the very uncertain foundation soil.

**Above:** Fabric installers make a final closure between fabric strips.

**Above:** Architect's concept sketch for the dome with masts and a series of big spheres around the perimeter which were changed to less expensive cylinders.

Millennium Dome

The structural concept of the roof is of tensioned radial stringer cables which support the fabric and run between the inner ring and the concave curve of the fabric edge. In each of the twelve segments making up the flat hemisphere the centrally located mast is positioned by a forestay tied to the inner ring and two backstays attached to ground anchors at the two bottom corners of the segment. The network of so-called hanger cables is also attached to the top of each 100 metre (323 foot) high mast and supports and shapes the fabric far below, along the lines of five intermediate circumferences around the dome. These circumferences are marked by cables located 2 metres (6.5 feet) above the fabric surface to avoid ponding of water or snow. Over such large distances the fabric appears to be curved but in fact, like a marquee tent, it is composed of flat panels. The individual fabric panels are wedge shaped and this has an added advantage: the panels deflect more as they get wider which has the effect of shedding water and snow in the right direction – downwards.

Following a change of government and a change to the expected life of the structure, the original PVC-coated polyester fabric was replaced by two layers of PTFE-coated glass fibre fabric.

The masts are all identical. They are formed from eight steel tubes 323 millimetres (13 inches) in diameter braced with rings every 2.5 metres (8 feet). They have to resist not only the normal buckling loads but snow and ice loadings as well. Like the whole structure they can survive the removal of one component – even the support tripods will still work despite the removal of a leg.

A special challenge was presented by the existence of a pair of major motor traffic tunnels under the Thames, which had implications for the foundations and also required a very large concrete vent above ground. The former problem was solved by using flight auger-style piles rather than the driven piles used elsewhere, while the vent was accommodated by a carefully engineered hole in the dome fabric.

Raising the structure took some management. With the masts supported by backstays and a temporary forestay, the central ring was raised using the spare permanent forestay. The cable net was then assembled and hoisted into place. Following tensioning of the radial cables the rest of the cable net was installed. And then the fabric was fixed.

**Right top to bottom:** The sequence of construction. First, with the masts temporarily stayed and the steel oculus suspended from them. Then, with some of the fabric in place and, later, completely clad. Finally, as an operational structure complete with service towers around the perimeter.

# Engineer: **Buro Happold**

Japan Pavilion, Expo 2000, Hanover, 2000

Architect: **Shigeru Ban with Frei Otto**

**The Hanover Expo 2000 had the environment as its main theme and Shigeru Ban's Japanese pavilion was designed to deploy materials which would produce the minimum of waste when it was dismantled. Ban decided that the structure should be of one of the most easily recyclable materials, paper. In fact the main structural members were cardboard tubes. Ban had first used these as a framework for plastic sheet and tarpaulin emergency houses during an aid programme in Rwanda, where they had been developed by the packaging company Sonoco Europa. Later Ban had used them on other projects, notably in his post-Kobe earthquake church, whose wall structure was in cardboard tube.**

Consultant to the project was the great German architectural engineer Frei Otto who brought in Buro Happold, the office of his old collaborator, the late Sir Edmund Happold. The engineers immediately began materials tests and stress analysis, the latter deploying an inverted model to which loads could be attached to replicate the full size conditions.

 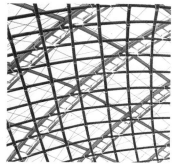  

**Above:** The undulating grid shell with its narrow transparent half rings and the surrounding timber fence.

**Opposite:** The three layers of the Expo pavilion. Bottom, the sand-filled scaffolding board boxes used as temporary foundations. Next, the cardboard tube grid shell with end walls of similar construction. Then, the array of ladder arches, originally having no structural function. Finally, the translucent skin, originally paper but eventually, courtesy of the city fathers, paper and transparent PVC.

Early ideas for the pavilion centred around the idea of a big 72 metre (236 foot) long tunnel made from paper tubes along the lines of an earlier design by Ban for a dome. This latter had wooden connections which had been expensive and Ban decided to bypass the problem of nodes when he realized that cardboard tubes can be made to any length and very long lengths may be curved. In fact, the tubes could be made 20 metres (66 feet) long and spigoted together to form continuous runs of as much as the 68 metre (223 foot) lengths used here. Ban suggested to Frei Otto that this was a better approach than using short struts with node connections because the tubes could simply be tied in their 1 metre (3 foot) grid using rope. The whole assemblage could be set out and tied up while it was laid out flat on the ground. Forklifts could then be used to push the structure up to its final arched profile. Ban says, 'We wanted to finish the entire structure using methods that were as low tech as possible, so we argued for a simple joint of fabric or metal tape.'

The original idea was for a double curvature grid shell. It was not a simple barrel vault, rather the final shape of three tightly integrated hemispheres with diameters of around 35 metres (115 feet) each and heights of more than 16 metres (52 feet). But buckling analysis showed that very big diameter tubes would be needed and they would be in danger of being damaged in the process. An existing element came to the rescue. The paper skin was to be supported on a 3 metre (10 foot) grid structure lifted clear to avoid trapping moisture between paper skin and cardboard tube. This structure was a series of timber ladders at 3 metre (10 foot) centres, arching across the vault with single timber purlins, also at 3 metre (10 foot) centres, running longitudinally and 8 millimetre (1/3 inch) stiffening cables underneath. Cladding is a double membrane of paper with a protective polymer-coated polyester over it and single clear polymer over the ladders.

Japan Pavilion, Expo 2000

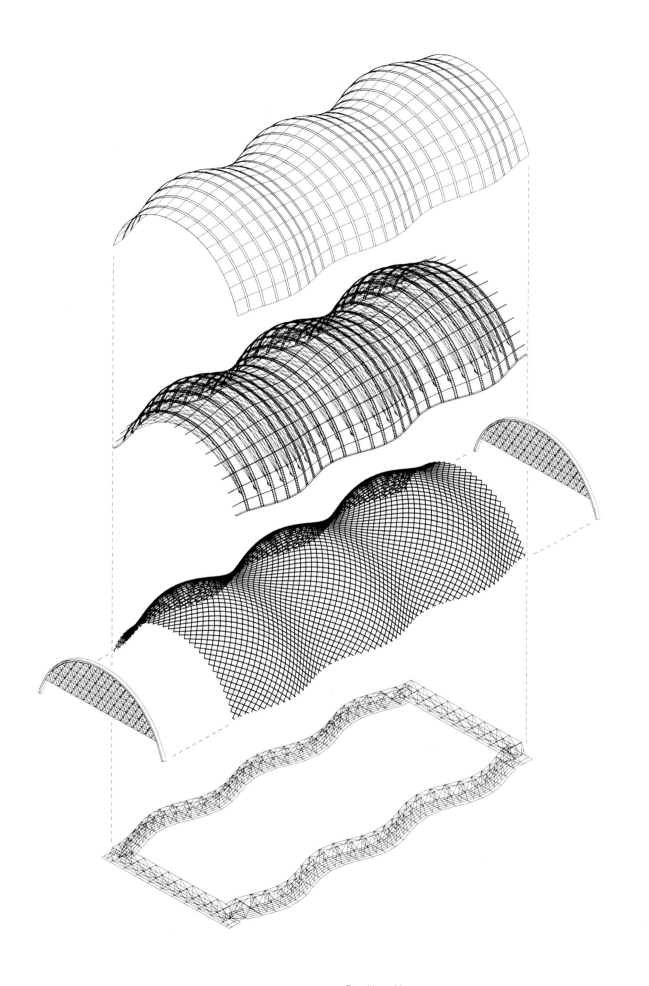

**Above:** The interior of the grid shell. Bizarrely the exhibit was given a false fabric ceiling so the extraordinary structure was understood only by people who knew about it.

**Below:** Ladder arches fixed over the structural grid shell and restrained laterally, with a section of the final skin in place.

Japan Pavilion, Expo 2000

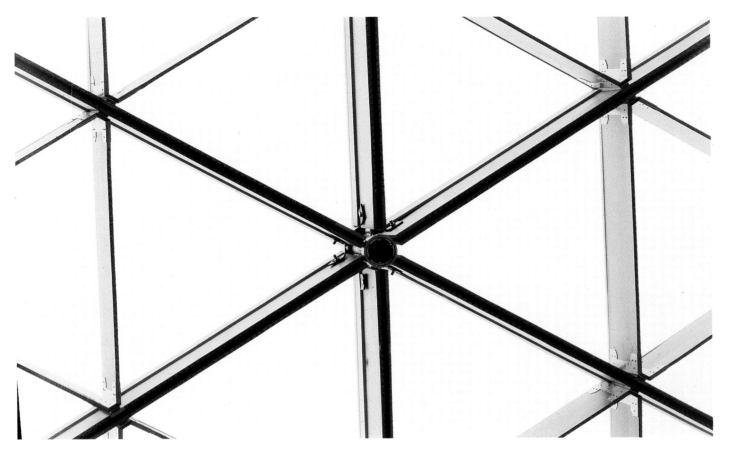

**Above:** Detail of the diagonal structure of the flat end walls.

**Below:** Plan of the pavilion.

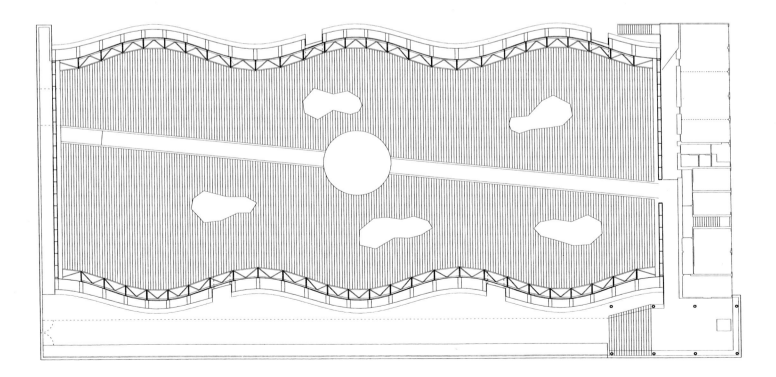

**Above:** The cardboard tubes of the grid shell are simply tied at their crossings with rope and, here and there, tied back to the ladder arches.

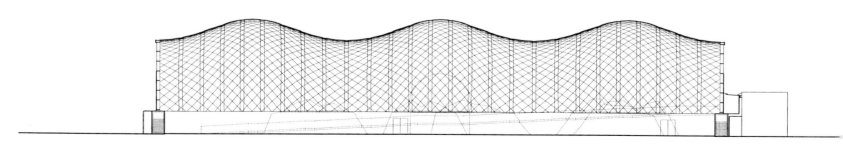

**Above:** Long elevation of the pavilion, which is actually sections of three spheres moulded together.

Japan Pavilion, Expo 2000

Ban describes the events thus: 'Buro Happold suddenly pointed out a major structural defect. There was an unexpectedly large of amount of creep in the paper tubes, which made it impossible to guarantee an adequate safety ratio for the grid shell. Buro Happold proposed several ideas as countermeasures, but at this point it was impossible to make any major changes to the form or function of the pavilion, or any changes that would significantly increase costs. The paper tubes were being tested, so their size could not be changed. Although it detracted from the purity of the paper tube architecture, we decided to combine the grid shell and timber arches by abolishing all of the joints and bracing cables in Buro Happold's original proposal for the grid shell and enlarging the sections of the timber rafters, which originally were to have served as a frame for the membrane and scaffolding for construction work.'

So the final structure had the 120 millimetre (5 inch) paper tube grid shell defining the form of the building. Over this is an array of ladder arches (to which the grid shell is tied), stiffened by underslung cables which do not have full structural continuity and provide out-of-plane buckling resistance. The cardboard tube grid was pushed into the final shape over three weeks using temporary scaffolding. When in place the timber ladder arches were put up and tied to the grid shell. Then the diagonal cables were lightly tightened up and the skin fixed. Once the centre six ladders were in place the builders worked out on either side to the ends. The ends of the great structure were closed with vertical diagonal cable nets. The nets, which provide stiffness as well, are tensioned between laminated timber end arches. They are based on 12 millimetre (1/2 inch) steel cables in a square grid, into which are fixed timber and paper honeycomb board composite panels covered with paper or PTFE glass membranes. The ladders and the paper tubes are anchored around the perimeter of the building by steel frames enclosing scaffolding board boxes loaded with sand rather than eco-unfriendly concrete.

During construction the city of Hanover authorities seem to have lost their nerve, insisting that Ban's engineer be replaced with a local one and demanding recalculation of the structure on the basis that it was rigid, rather than flexible. This meant that timber sections had to be increased and steel reinforcement added. Ban continues, 'Then, finally, an incredible restriction was placed on the push-up construction method, limiting it to 2 centimetres [3/4 inch] per day instead of the planned 20 centimetres [8 inches]. This would make it impossible to meet the May opening schedule. Because of this last impossible demand, we made the heart-breaking decision to accept all of the city's requirements. But this was not the end. Although the roof and paper membranes had already cleared the fire standard tests, on the pretext that the pavilion might become a target for terrorists, the demand was made that we replace the roof and paper membranes with conventional PVC membranes rated one grade higher on the fireproof scale. We could not accept abandoning the paper membrane developed for this project, so we decided to place a transparent PVC membrane above the paper membrane.

'I was surprised by the reluctance of the city bureaucracy to recognize new structures and new materials, and above all by its unwillingness to listen to an authority of Otto's stature and achievements. Without Otto's co-operation, an advance such as paper architecture would probably have been impossible.'

While the Hanover pavilion was being erected Ban and Happold were putting up a similar temporary grid shell, this time an open structure over the sculpture court of the Museum of Modern Art in New York. Weighing around 9 tonnes (9.84 tons), it was 9 metres (30 feet) high and spanned the 26.5 metre (87 foot) width of the garden until it was dismantled with the beginning of the museum's major refurbishment programme.

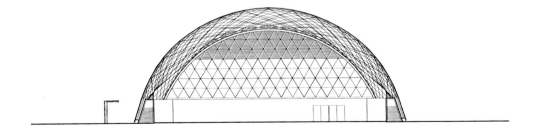

**Above:** End elevation of the pavilion showing the extent of its diagonally patterned structure.

# Engineer: **Anthony Hunt Associates**

## Eden Project, Bodelva, Cornwall, 2000

## Architect: **Nicholas Grimshaw & Partners**

**One of the Millennium projects funded by the national lottery, the Eden Project is located in an old china clay quarry in Cornwall, on the south-west peninsula of England. The irregular succession of eight immense, interconnected hemispheres – christened 'biomes' – form two groups half buried in the quarry wall and are connected by a restaurant and plant holding area. This is the world's largest artificial plant environment, able to accommodate anything from large trees to desert rock plants in a variety of zoned environments which are maintained, as far as possible, by sustainable technology. In a similar vein, part of the roof of the link building containing the restaurant is covered with turf.**

Until the initiation of the Eden Project, the site was used as a clay pit. This created a problem of ground levels for the structure and led to the decision to deploy the hemispherical geodesic forms which had been popularized in the 1950s by the American visionary, Buckminster Fuller.

A critical item in the client's brief was that whatever the form of the enclosing structure, it should have very high levels of light transmission. This implied that glass, or something similar, was needed and that the supporting structure should be as slender and therefore lightweight as possible. The engineers and architects looked at a long list of possible forms and finally agreed on a single layer, unbraced, three dimensional space-frame system.

**Preceding page:** The Eden Project skin, extraordinary delicacy and giant scale.

**Above:** Computer image of the biomes nestling into the lee of the former quarry, with the visitor centre in the foreground nudging into the opposite quarry wall.

The German firm Mero won the contract with a more traditional double layer space truss system. Its outer layer took the form of giant hexagons 5 to 11 metres (16 to 36 feet) in diameter, while the semi-braced inner layer was a combination of hexagons, pentagons and triangular elements known as a hex-tri-hex arrangement. Although this configuration weighed considerably less, it involved a large number of expensive nodes and made the prefabrication more complicated, so it was best to keep the number of nodes to a minimum by maximizing the size of the hexagons. Glass glazing would have required smaller hexagons and considerable loadings, as well as maintenance and replacement problems. The design team instead decided to deploy a pneumatic structure of pillows formed from multiple layers of ETFE (ethyltetrafluoroethylene).

This is a modified copolymer extruded into an extremely thin film which gives very high levels of light transmission in both the visible and ultra violet range. It is one-hundredth the weight of glass, but has as good an insulation performance as double glazing. Since its surface is extremely smooth and has anti-adhesive properties ETFE is self cleaning. Unaffected by ultra violet light, atmospheric pollution and weathering, its anticipated life – partly based on its performance in Burgers Zoo in Arnhem, Holland – is in excess of 40 years. It then becomes biodegradable.

The ETFE cushions are zipped into aluminium frames, using a system rather like a luff rope on a sailing craft, and follow the mostly hexagonal geometry of the outer layer. The pillows are usually hexagonal but in some cases, where the geometry of the biomes changes, they are triangular or pentagonal. Since the material is so light and has good structural performance, some hexagonal pillows are up to 11 metres (36 feet) across but weigh only 50 kilos (110 pounds).

Air pumps keep the ETFE pillows inflated. It was desirable to make the pillows as large as possible, but above a side length around 5.5 metres (18 feet) they needed reinforcement. So the engineers had to keep the hexagons big to minimize structure but not so big as to require complicated pillow reinforcement. In the final structural equation only some of the very largest pillows were reinforced.

The combined dead weight of the structure and ETFE glazing is naturally very light, around 40kg/m$^2$ (8lbs/1ft$^2$), so live loadings of snow, wind and so on seemed significant. Snow loads were important because there was the probability that drifts would collect in the valleys between the biomes and the spaces between the pillows. This was calculable, but it was impossible to estimate wind loads from the structural building codes. It appeared that because the top of the biggest biome would be 10 metres (33 feet) below the top of the clay pit edge it could be thought of as a below-ground structure. Wind tunnel testing confirmed this and, happily, indicated that the engineers had been very conservative in their estimates of the effect of wind.

**Below:** Still under construction. The site is about to be landscaped and the remaining pillows zipped into position on their complex structural support.

Because the size of pillows was optimized for each hemisphere their junctions presented problems. In the end the simple solution was adopted – tubular lattice arches to which the perimeter node points of adjacent hemispheres were simply attached.

In this project each decision about materials, loads and shape had implications on everything else, so there had to be total co-operation between the architects, engineers and main contractor. The primary vehicle of this collaboration was a three dimensional digital model of the design which was passed electronically between the major people in the combined team.

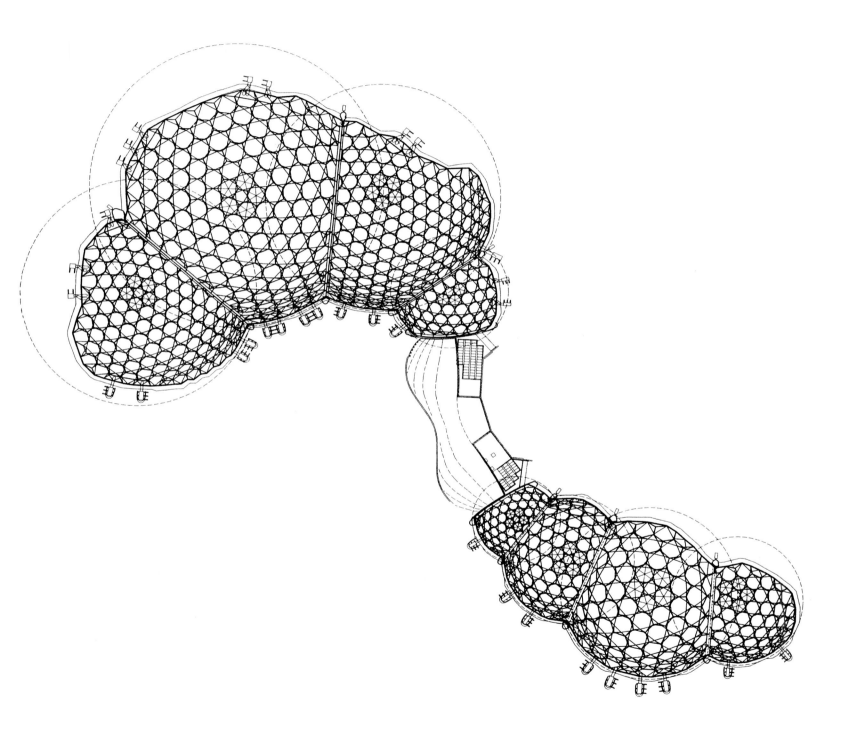

**Above:** Plan of the Eden Project, with four biomes of unequal sizes linked by a grass-roofed restaurant and plant holding areas.

Eden Project

**Above:** The extraordinary, ethereal quality of the domes and their supports, more like a network of veins than a structure.

**Below:** The visitor centre is a major fabric structure in its own right.

**Above:** Originally conceived as a single layer geodesic structure, the geometry was changed to this more complex but lighter, double layer hex-tn-hex configuration. It has an outer layer of hexagons and an inner layer of triangles – mainly tension cables.

**Below:** Engineering detail of a node. The tubes are sides of three adjacent hexagons. The double cables are part of the lower structural layer. The extruded strips on top of the tubes are grooved to take the edges of the ETFE pillows.

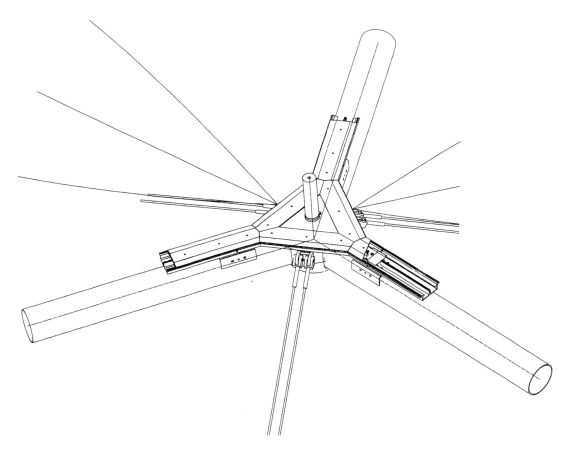

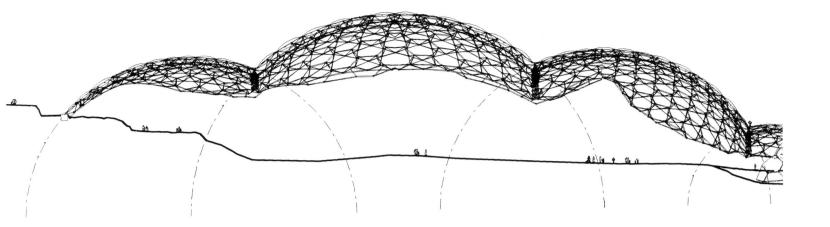

**Above and below:** The geometry of the biomes is based on interconnected spheres but the structure is not unyielding because it is easily adjusted to cope with local changes in topography.

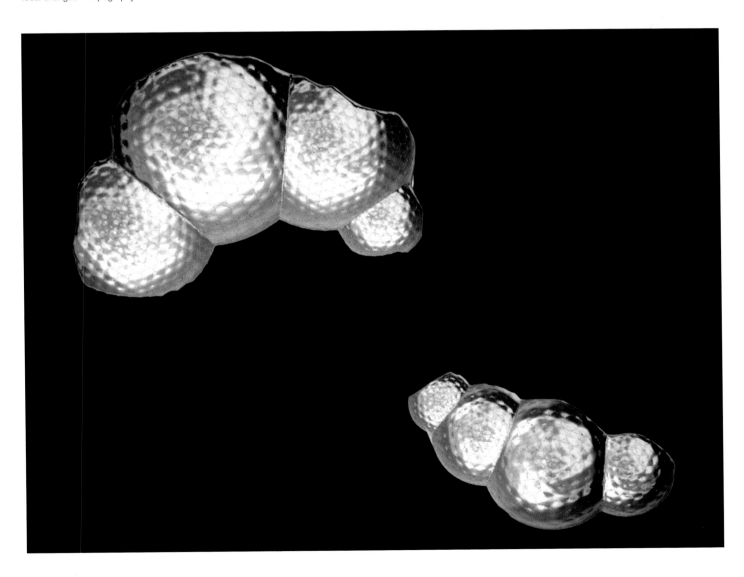

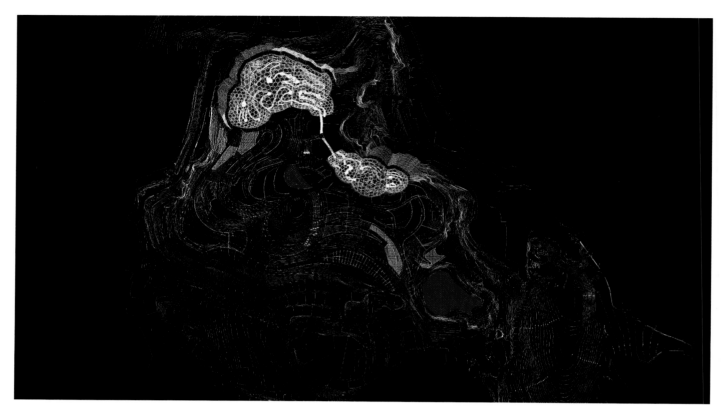

**Above and below:** The organic shapes of the biomes and their unusual setting in a vast hole in the ground have led to some dramatic computer images.

**Opposite:** The fabric roof of the visitor centre anchors directly into the grassy bank at its rear.

Eden Project

# Engineer: **Anthony Hunt Associates**

Great Glasshouse, National Botanic Garden of Wales, Carmarthenshire, 2000

## Architect: **Foster & Partners**

**This great elliptical dome is the world's largest single span glass vault. It creates the temperate environment for a vast collection of plants and fungi and for a hands-on learning centre.**

**It is a symmetrical shape, but one side is canted up so the axis of the short cross section is at an angle and the vault is inclined 7 degrees to the south and the sun. Despite its size the structure is remarkably sparse: tubular steel arches are seated on a tilted concrete ring beam to which the glazing system is attached. Slots cut in the encircling earth berm give access under the high edge of the vault to the interior. Some glass windows pivot up and down automatically as part of the environmental controls.**

Engineer Anthony Hunt made an intensive study of geometric form that approximated to the doubly curved oval shape desired by the architects. The geometry finally selected was based on a section from the outer side of a torus, a doughnut-shaped form created by the rotation of a sphere about an external axis. The use of a section of this regular figure is reminiscent of the Utzon/Arup decision to use segments of a sphere for the Sydney Opera House sails. It also had the manufacturing advantage that all the curved beams, which range in length from 58 metres (190 feet) at the widest point, have the same radius because they follow the geometry of the toroid (see page 65).

Simple structures rarely engender simple engineering problems and this concept called for advanced non-linear buckling analysis. Canting the dome introduced extra bending stresses in the beams so additional 150 x 248 millimetre (6 x 9 3/4 inch) T sections, part of the glazing support system, are continuously welded to the tops of the 324 millimetre (12 3/4 inch) diameter tube arches. These provide extra stiffness because tube and T section now act as a composite beam. Bracing against lateral buckling takes the form of 140 millimetre (5 1/2 inch) diameter longitudinal tubes running longitudinally at 7.5 metre (25 foot) centres.

The main arches were brought on site in several sections, then welded together and the ball joint at the tapered ends welded on. The ball for each tube arch sits in a socket welded to the steel base plate bolted on the concrete ring beam. This system meant there was no need for the potentially complex alignment that a more fixed system would have

occasioned during erection. It was discovered that under extreme wind loads the arches at the ends have to resist uplift, so here there is an additional capping piece around the ball and socket which is bolted down to the base plate.

The glazing system is an anodized aluminium glazing frame incorporating a gutter system. The water drains in channels that run down the shorter sides of the vault and are fed by shallower horizontal gutters, so the grid of the glazing system is also the integral grid of guttering. The aluminium system is free to move a certain amount in its fixings to the steel frame, so eliminating any problems of differential thermal expansion. The glazing is 17 millimetre (3/4 inch) laminated clear glass, silicone bonded to the aluminium frame except where the lights open and the glass is fixed in an aluminium window frame hinged to the main framing.

**Preceding page:** The geometry of the vast glasshouse is taken from the outer edge of a torus.

**Above:** The vast glazed shell nestles into the crown of the hill.

Great Glasshouse, National Botanic Garden of Wales

**Horizontal or tilted**

**Outward thrust to be revisited**

**Ground profile is independent of edge condition**

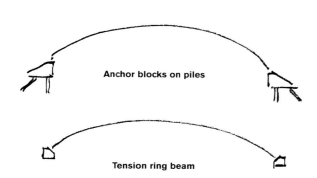

**Anchor blocks on piles**

**Tension ring beam**

**Above:** Engineer's drawings exploring possible configurations and structural principles.

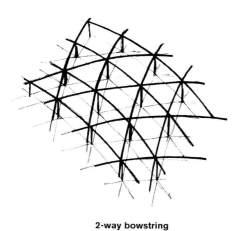

**2-way bowstring**

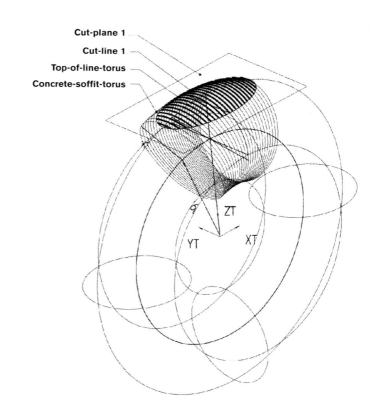

Cut-plane 1
Cut-line 1
Top-of-line-torus
Concrete-soffit-torus

ZT
YT  XT

**Above left:** One early possibility was to deploy a two-way bowstring grid.

**Above right:** The geometry based on a torus.

**Above:** The warped grid is that of the glazing system rather than the structure, which is a series of big tubular arches.

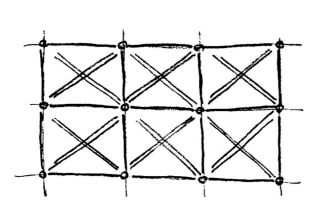

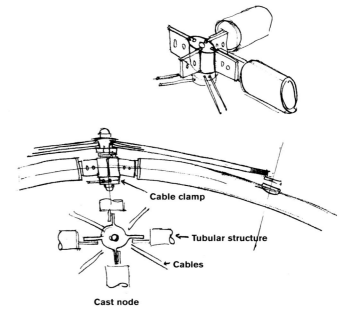

**Cable clamp**

**Tubular structure**

**Cables**

**Cast node**

**Above left and right:** Another possibility was to use a single layer two-way grid dome for which the engineers had worked out preliminary junction details.

**Below:** The tubular arches are seated on ball joints. At the ends they have restraints to stop the arches pulling up. The glazing system is fixed independently on top of this basic arched structural system.

# Engineer: **Ove Arup & Partners**

## William Hutton Younger Dynamic Earth Centre, Edinburgh, 1999

## Architect: **Michael Hopkins & Partners**

**The Dynamic Earth Centre is located off the Royal Mile, on the site of the home of the eighteenth century father of geology, William Hutton. The Centre is an exposition of the forces and events which have created and moulded the earth. The Younger in its name is that of the company which donated this former brewery works site to the city of Edinburgh.**

**There are four elements in this very formal composition. One is the monumental, amphitheatre-like forecourt; the second, at its top, is the great fabric canopy which serves as a rooftop pavilion and entrance to the third element, the half-buried two storey exhibition block below; the fourth is the underground car park which extends under the whole site. The great igneous rocks of Arthur's Seat and Salisbury Crags form a looming background.**

The exhibition space is essentially a big, two-level black box (with administration offices at the sides), some of whose boundaries are defined by fragments of the old brewery building. The plan of the glass walled, fabric roofed pavilion atop is a big oval. The hemispherical roof of the black box's multimedia theatre bursts through the centre of the floor from the level below. The drama of this intrusion is enhanced by latitude and longitude lines, as if the hemisphere were tilted to echo the dynamic character of the roof above, with its raking support pylons. At the focal points of the end semicircles of this great, unheated space are circular islands accommodating basic catering facilities and stairs and lifts down to the exhibition.

The Hopkins office has designed a number of other significant fabric structures with engineer John Thornton of Ove Arup & Partners. Here they made the roof a symmetrical, segmented, vaguely beetle-like shape, with four paired masts emerging from four glazed ladder girder saddles across the PTFE-coated glass fibre fabric. The glass is toughened safety glass and point fixed to the ladders. In the middle of the wall overlooking the 'amphitheatre' the edge of the canopy is lifted up and extended to form a kind of porte-cochère propped up with two diagonal struts. The main masts are reciprocally stayed. Tapered at top and bottom, they carry the ladders via steel rods attached around the mast some distance above and below the line of the ladder. Diagonal lateral cables connect from the same position to locate the underside of the intermediate 'coat hangers', the curved steel tube girders suspended from the tops of the masts which support the roof's three intermediate ridges.

Around the edge of the fabric the anchors are kept close to the perimeter of the roof by using an angled compression strut that allows tension cables (actually steel rods) to run at a steep angle to the ground. The alternative would have been a series of cables following the line of the roof profile to anchors well outside the edge of the black box below. Structural logic would then in some cases have called either for a larger base structure or a smaller canopy. As it was, at the west end the curving end of the canopy is dealt with by using two long compression props and anchoring the cables to the wall of the exhibition box via an outrigger. The glass wall is an independent cantilever structure. Its top perimeter is flexibly connected to the underside of the main fabric roof with a loose air seal flap to allow for movement in the fabric.

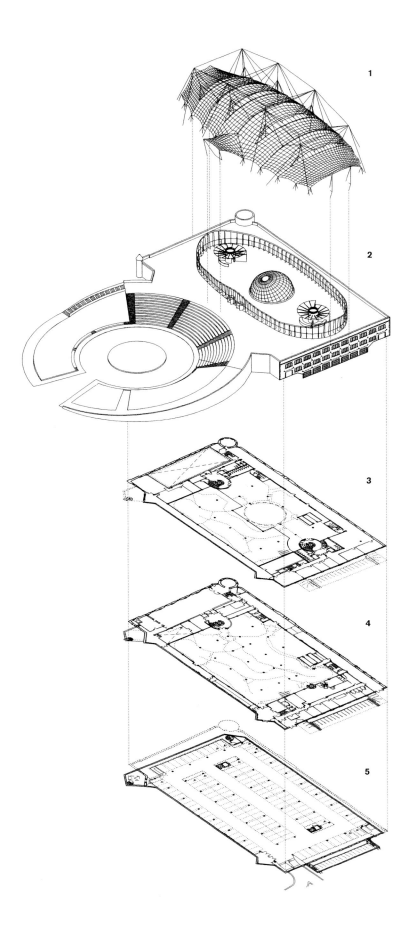

**Preceding page:** Under the great translucent canopy the embedded tilted hemisphere promises the answers to secrets beneath.

**Right:** The five layers of the centre. **1.** Fabric roof structure including struts. **2.** Entrance floor and open air amphitheatre. **3. and 4.** Exhibition space. **5.** Car park.

William Hutton Younger Dynamic Earth Centre

**Above:** The bulk of the exhibition space lies behind the curving steps of the open air auditorium backed by an idiosyncratic yet delicate tent.

**Below:** At night the formal arrangement lends itself to dramatic lighting.

**Overleaf left:** Props at the perimeter of the fabric roof enable the loads to be taken at a steep angle to keep within the curtilage of the building.

**Overleaf right:** Always a problem with fabric roofs, the junction between roof and wall is carried off by a flexible fabric flap. This connects loosely to the top of the all-glass wall, which remains erect with the help of glass fins cantilevering up from the floor.

William Hutton Younger Dynamic Earth Centre

Ove Arup & Partners

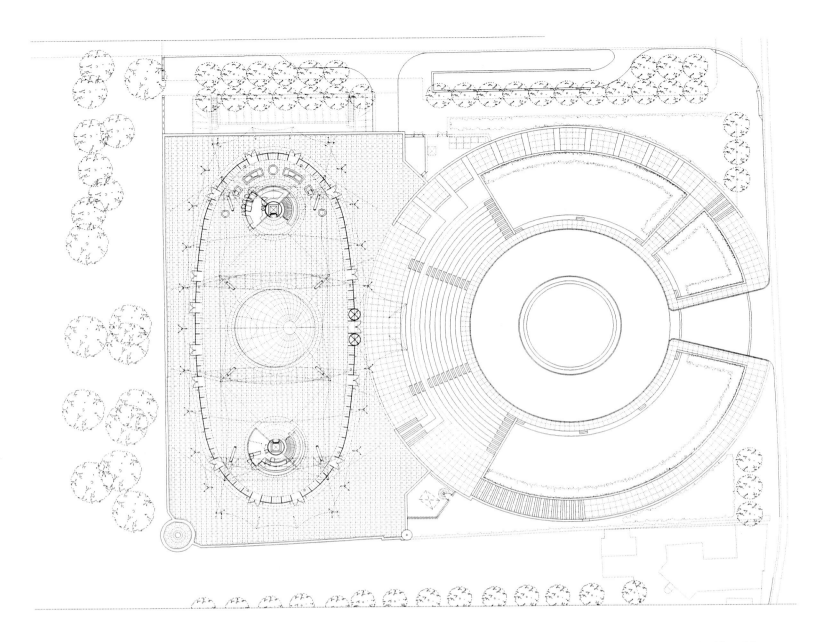

**Above:** Ground plan of the site.

**Below:** Section across the building with the open air auditorium on the right and the fabric roof over the exhibition space and car park.

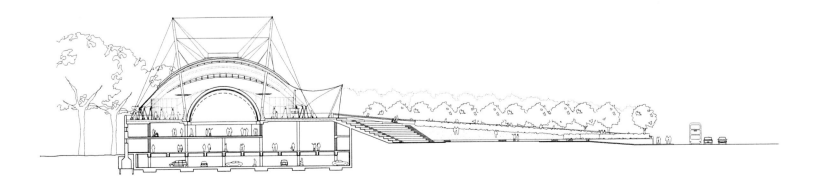

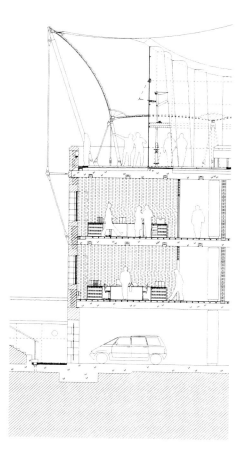

**Above left:** High technology glass and high technology fabric engineering.

**Below:** The lightweight fabric structure nestles at the foot of the granite hills.

**Above right:** An extension of the propping devices is used to carry diagonal loads at the edge of the fabric roof down vertically to keep within the perimeter of the building.

# Engineer: **Weidlinger Associates**

Rose Center for Earth and Space, American Museum of Natural History, New York, 1999

## Architect: **Polshek Partnership Architects**

**The trustees of the American Museum of Natural History were bemused when their architects, James Polshek and Todd Schliemann of Polshek Partnership, put forward their proposals for replacing the outdated 1935 Hayden Planetarium at the side of the museum's site on Central Park West between West 77th and 81st streets. The architects offered them the somewhat hermetic idea of a sphere within a glass cube: the sphere and cube being two of the small number of perfect, three dimensional geometric forms.**

They have a certain resonance with the designs of the late eighteenth century symbolic French architects, among them Ledoux, Boullée and Le Queu, while in Platonic theory these forms are the symbols for two of the four elements, air and earth. The modern proposition might be that the building's form represents the sphere of the cosmos enclosed transparently by the cube of earthly comprehension.

PHINEAS & SANDRA PRIEST ROSE CENTER FOR EARTH

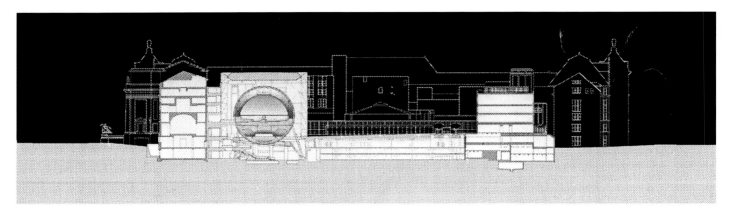

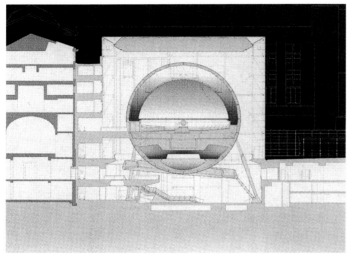

The upper half of the sphere contains the Space Theater, a planetarium with a suspended hemispherical ceiling on which are projected various sky maps and displays. It is accessed via a gangway from an upper floor of the old museum building which forms one of the back walls. The lower half of the sphere contains the Big Bang Theater and has an entrance next to the beginning of the spiral walkway.

The main engineering issues were the organization and support of the sphere in space, the design of the daring engineering of the 111 metre (364 foot) spiral walkway and the creation of the great glass walls on the north and west façades of the new cube. These were conceived as single vast panes of glass rather than conventionally framed curtain walling.

**Top:** Section through the museum with Central Park to the left.

**Above:** The sphere in more detail showing some support struts and the spiral ramp at the bottom.

Early schemes proposed by architect and engineers had the 26.5 metre (87 foot) diameter sphere sitting on a variety of alternative supports. The trustees finally approved a design in which the sphere appears to float in a well in the cube's floor, with a free-standing spiral walkway winding around its nether regions and leading down to the museum's lower floor. Despite its apparent buoyancy, the sphere is of course supported in its 37 metre (120 foot) square glass enclosure. Three clusters of props based in various locations on the lower floor support not the base of the sphere, as one might expect, but random points around its equator. The impression of floating is due partly to the fact that the bottom of the sphere is thus clear of support, and partly because the props are slender and lack the orderly structural arrangement one expects of engineering.

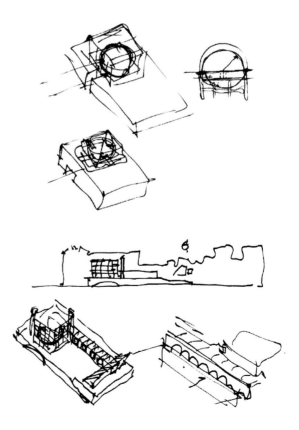

**Preceding page:** Conscious references to visionary French eighteenth-century architects Ledoux and Boullée – here two centuries later a reality.

**Above:** Early architect's concept sketches including ways of supporting the sphere.

**Opposite:** Construction of the lower floor and spiral ramp, before commencement of work on the sphere whose lower section hovers above this spiral.

Rose Center for Earth and Space, American Museum of Natural History

Frameless glazing was a kind of ideal for twentieth century architects. It is only recently that it has been possible to achieve something like it, and then using glass considerably thicker than the 12 millimetre ($^1/_2$ inch) low iron, monolithic glass of the Rose Center. Weidlinger Associates created a structure of vertical tube steel trusses to which are fixed the smaller elements making up the attachment system. There are also three intermediate horizontal tubes at 6 metre (20 foot) centres behind the two glass façades. Each of the 3 x 1.5 metre (10 x 5 foot) sheets of glass has a spider attachment at each corner locating it 10 millimetres (1/3 inch) away from the adjacent three sheets, the gap filled by silicone sealant. The spiders are located in space by a system of vertical and horizontal tension rods and stand-offs. At ground level this is substituted by stabilizing glass fins. The glass folds over the top of the structure creating a glazed band around the edge of the opaque roof. Most of the panels on the west façade are given a 50 per cent fritting in a pale tint to diminish the effects of the western sun.

The sphere has an easily understood structure. It is effectively a circular platform supported by angled struts, with the Big Bang Theater suspended below and the whole encased in a structure of curved steel ribs clad with aluminium panels. These panels were computer-cut, then stretch-formed along two axes as segments of a sphere and given an internal coating of a non-woven acoustic fabric. The ramp is cantilevered from a 750 millimetre (30 inch) wide torsion tube supported every 27 metres (90 feet) by angled props. In a complicated geometry it spirals inwards as well as down.

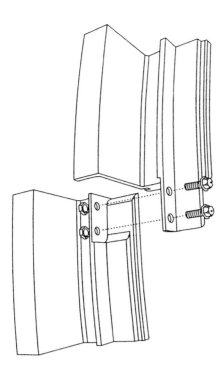

**Above left:** Detail of the secondary system to which the sphere's cladding is attached.

**Above right:** The ribs of the sphere are connected together.

**Below:** The basic steel structure of the intermediate floor for the planetarium – which is also the ceiling for the Big Bang theatre slung underneath in the lower hemisphere.

Rose Center for Earth and Space, American Museum of Natural History

**Below:** The immensely elegant secondary structural system in which the offset vertical rods work with the glass as quasi bowstring trusses.

**Below left:** The primary structure's trusses rise vertically while two of the sphere's props emerge from the rectangular well in which the sphere seems to hover.

**Below right:** The glass wall's vertical trusses have pin joint connections to supports. In between, the lower section of the glass wall is supported by glass fins – also mounted with pin joints so the whole structure can move to accommodate and dissipate stress resulting from loadings.

# Engineer: **Structural Design Group**

## Tokyo International Forum, Tokyo, 1996

## Architect: **Rafael Viñoly**

**This vast building occupies the 2.7 hectare (6.5 acre) site of the former Tokyo city hall, not far from the Ginza commercial and pleasure district, and faces east to the outer gardens of the Imperial Palace and west over railway tracks. To the north and south are underground railway systems, each with a station.**

Commissioned by the Tokyo city council, the forum was conceived as a centre for cultural exchange with performance spaces, offices, meeting rooms and receptions. The design, by Rafael Viñoly of New York, is based on a row of four cubic masses containing big auditoria which are connected to the adjacent lobby by overhead walkways. The lobby is actually a seven storey glass hall with an unusual plan based on a narrow, sharp ended ellipse, 210 metres (689 feet) long. A range of walkways attached to the inside of the building's curving skins act as stiffening 'ribs'. Flying gangways sail diagonally and vertiginously across the great space, serving also as internal bracing to the whole structure.

   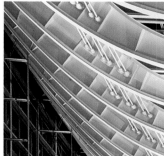

At ground level is a big exhibition area. Overhead light filters through from the upper section of the great hall and through its glazed roof. Sixty five metres (213 feet) above, this slightly curved glazed roof is supported by an astonishing structure in which the great steel trusses that both separate and connect the two sides of the hall appear to be upside down arches. This is a reasonable assumption for in a conventional structure, glass roofed or not, one expects the roof beams to span across the shortest distance between the supporting walls. Here however the whole structure is dependent on an immensely complicated 210 metre (689 foot) long truss supported at each end by a giant column. The upside down arches have indeed a stiffening function, but they serve to locate the longitudinal tension cables forming the lower chord of the truss – as well as the great compression longerons running the length of the roof and wishboning together over the support column at each end. So what appear to be inverted arches are actually skeletal ribs for a truss literally in the shape of a three dimensional bending moment diagram.

The glazed skin has no supporting function. It is effectively hung from the ends of the 'arches', then secured around the perimeter of the great hall and given stiffness against wind loads by vertical quasi-bowstring trusses. These trusses are spaced 10 metres (33 feet) apart. Their compression members comprise the glazing mullion and the short struts attached at right angles to it, which locate the lower chord of one cable and a pair of curving cables forming opposed bending moment diagrams. This is a tour de force in the expression of not so much function as the very shape and form of the internal stresses and forces which enable the building's structural integrity.

In a country prone to regular and severe earthquakes Japanese structural engineers have to design with high safety factors – and the tendency is, naturally, to over-size structural members and to generally design ultra-conservatively. This makes the engineer's long history of daring structure all the more impressive.

**Preceding page:** The extraordinary truss of the Forum from which the walls are suspended.

**Below:** The roof plan shows the four auditoria, the railway track sweeping around the bottom of the site and the enormous curved lobby whose primary function is to serve the auditoria.

**Opposite:** The extraordinary Viñoly/SDG roof truss is supported at each end by an equally extraordinary column. The side walls with their tension structure stiffening hang from the edge of the roof. Diagonal walkways provide lateral stiffening.

Tokyo International Forum

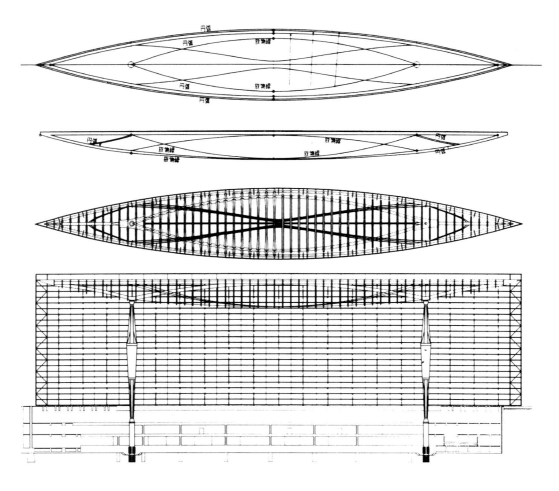

**Above:** Plans and sections showing the main elements of the great roof truss.

**Below left:** What look like inverted arches are actually unusual struts doubling as formers which locate the tension cables in their correct lines.

**Below right:** One of the two great columns supporting each end of the roof truss.

**Above:** Although it has the appearance of a purposeful structure, the great glass gallery is actually a lobby for the square auditorium blocks on the right and behind.

**Below left and right:** Early sketches showing that the basic design ideas were carried through.

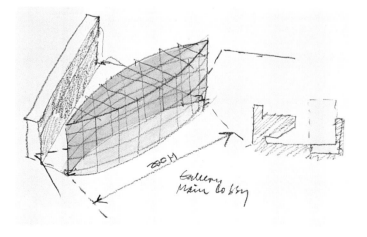

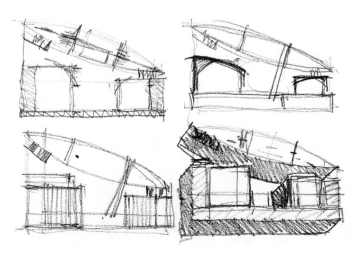

# Engineer: **Ove Arup & Partners** **with Braschel & Partner**

Mercedes Benz Design Centre, Sindelfingen, 1998

## Architect: **Renzo Piano Building Workshop**

**The model for this design facility for the German motor giant was the open human hand, the splayed fingers containing various activities of the company's design division which, inevitably, involve very high security. As they radiate out in increments of 9 degrees, the nearly overlapping roofs increase in length from south to north. These curving steel roofs do not quite overlap but lift up along one side to allow a continuous band of north clerestory light above the long wall of each finger.**

Three of the shorter roofs have short transparent sections maintaining the roof profile at their eastern ends and the area under them is cut out as a big semi-public exhibition space for showing off new models. The high walls between fingers are concrete and on the north and south, where they are end walls, they have an external skin of aluminium panels fixed to a polyethylene substrate which keeps the panels absolutely flat.

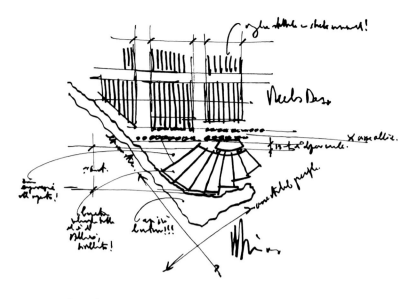

Renzo Piano self-deprecatingly puts it that this is a simple factory made up of long sheds. But the shapes are based on subtle curves and slopes initially worked out in conjunction with Jane Wernick and the late Peter Rice, then both with Arup. The clerestory lights increase in depth towards the open ends of the fingers as the separating concrete walls diminish slightly in size. The geometry of the roofs is taken from an uncommon, but well understood geometric form, the torus. Rice and Piano deployed the torus in developing the roof forms for the 1994 Kansai air terminal in Japan and Anthony Hunt used a different toriodal section for the National Botanic Garden of Wales structure (see pages 62–7). Here, in the Mercedes Benz building, the design team had taken a trapezoidal section of the outer surface of the torus. Its curve longitudinally is part of a big circle and its lateral curve is part of a much smaller circle, effectively the cross section of the 'doughnut' ring. Using regular, if uncommon, geometric forms reduces the amount of laborious computation because the primitives, the geometric shapes on which they are based, are in the structural engineers' central library of known forms whose properties and performance are thoroughly documented.

**Preceding page:** The great light of the Mercedes Benz design centre exhibition zone.

**Above:** Sketch plan of site.

**Below:** Translucent light baffles in the exhibition area.

Mercedes Benz Design Centre

**Above:** The roofs overlap, the upper edge supported on a glazed truss structure. As the diagram below right shows, the metal roof itself acts as the upper chord of a bowstring truss.

**Following pages:** Aerial view of the centre.

Maintaining the pure shape, which is defined by a material skin – here a sandwich of sheet steel with a thick insulating filling – involved a quite complicated structure. Two long tubes run longitudinally near the apex of the shallow curve of each roof and are connected to a series of struts in the form of inverted pyramids. Although at first glance these appear to make up part of a long space-frame truss, they do not. There is no lower longitudinal chord and the two beams have some supporting function; they also locate the inverted pyramids, which are there as part of a series of lateral trusses spanning between the top of the concrete wall on one side and the upper junction of the zigzag clerestory structure on the other. Each insubstantial truss is formed from the lower skin of the roof, the pyramidal, quasi-kingpost struts and a tension cable stretching from side wall to the top of the clerestory structure.

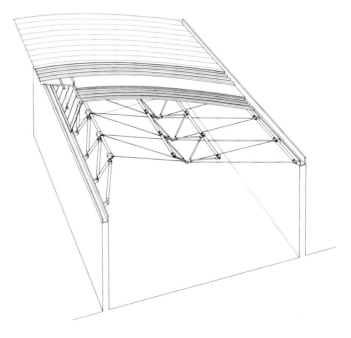

Mercedes Benz Design Centre

Ove Arup & Partners with Braschel & Partner

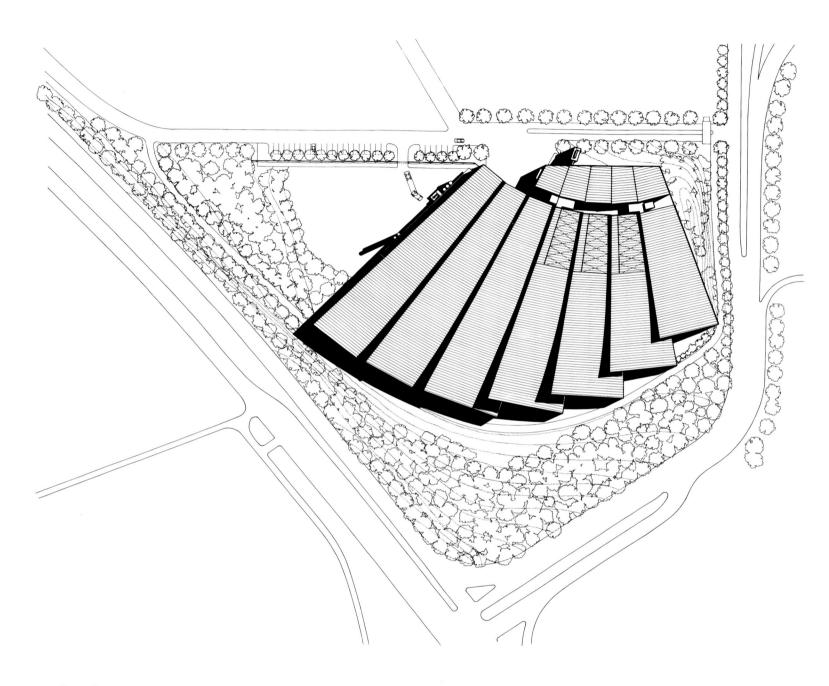

**Above:** Site plan with the exhibition space at the rear of the second, third and fourth
roof from the right.

Mercedes Benz Design Centre

**Above:** End view of one of the sections.

**Below left:** A torus showing the sections used for the Mercedes Benz roofs.

**Below right:** Using a section of a known geometric form made calculations much more straightforward.

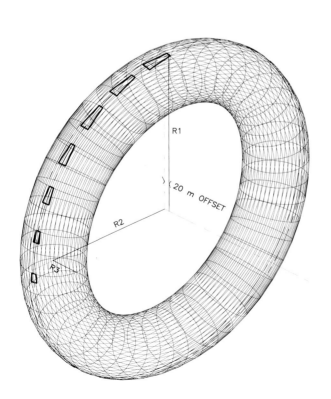

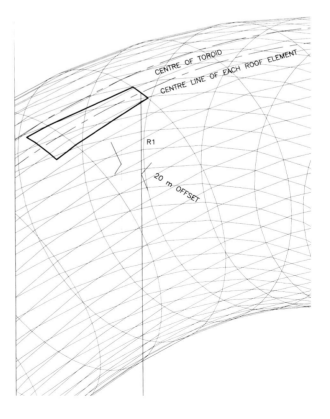

# Engineer: **Ove Arup & Partners**

Lords Media Centre, London, 1999

## Architect: **Future Systems**

**Arup engineer David Glover became involved in this project after Future Systems had won the competition with a concept design on which another Arup group had worked. The project design team's first task was to make a choice between the aluminium monocoque structure of the original proposal, in which the 'skin' of the structure is the critical structural element, or a more conventional steel structure clad with an aluminium rain screen.**

At the very beginning the team called in the boat building company Pendennis for expert advice on monocoques. Glover has written, 'We had to balance our knowledge of building with their knowledge of ship building and somehow bring out a tender that was workable.' They used a 35 metre (115 foot) motor cruiser, then being built by Pendennis, as guide for costs and volume of materials. A monocoque structure presents thermal and acoustic problems, plus the aluminium would probably melt in a fire. Solutions were found for all these problems and on paper the monocoque was only around 3 per cent more expensive.

Engineers operate within the limits set down by construction codes. An important part of Arup's work was to analyze existing codes for analogous structural conditions. They then had to go back to the code writers to discuss how the figures could be interpreted in the case of a non-rectilinear structure in aluminium, almost never used in the building industry as a structural material.

The basic structure on which the design team decided consists of two big, GRP-clad concrete legs, which support the aluminium shell and contain stairs, lifts and services. They are actually the upper part of two tension piles, extending 26 metres (85 feet) below ground level and around 12 metres (39 feet) above. The steel staircase inside each leg acts in conjunction with the reinforced concrete as part of the structure. On top of the legs is an integral concrete ring beam on which sits the shell, 40 metres (131 feet) from side to side and 20 metres (65 feet) deep. It is an odd structure because the load is eccentric, loaded towards the front where it overhangs the spectator stands.

The shell is not a true monocoque because it is supported at only two points and the shape, although it is heavily curved at the sides, has a relatively flat roof and base which act a little more like beams than structural shells. It is closer to a boat, with ribs and longerons and an integral waterproof skin which interacts with the rest of the structure – contributing 20 per cent towards the building's strength. So this is actually a semi-monocoque. However it does work as a single structural entity: there are no thermal expansion joints and when the white coated structure heats up the whole shell swells around a notional centre point. The building had to be white to keep heat expansion within sensible limits.

The monocoque was fabricated by a Dutch sub-contractor. The shaped pieces were first stitch-welded together and finally seam welded and the welds were then waterproofed. All but two sections of the shell were less than 3 metres (10 feet) wide, so they could be delivered without special police escorts to the site where they were assembled and welded together.

**Preceding page:** An unexpected sleek skin for a bastion of British conservatism.

**Below:** Press desks facing west through the canted glass wall overlooking the Lords cricket ground.

Lords Media Centre

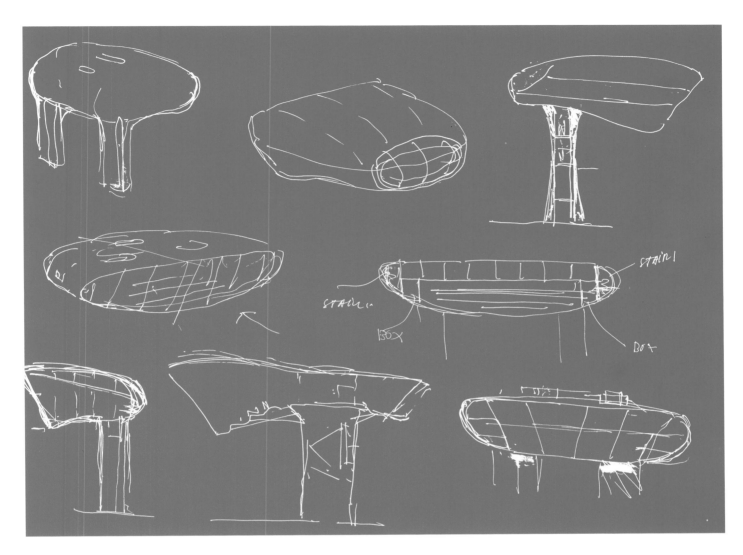

**Above:** Early concept sketches.

David Glover, the engineer, is adamant that the design is not wilful: 'I fundamentally disagree with that because I think it is a rational response to the problems that were set out in the brief. The beautiful process was to do with understanding the concept and the fact that the whole shape came from the brief. For me as an engineer that's wonderful.' Yet architect Jan Kaplicky points out that however advanced the building is in terms of form, shape, materials and structural concept, there is actually no new technology here – only technology new to the building industry. As his partner Amanda Levete says, 'We battled very hard to persuade them [the client] to use the boat builders. If we'd lost that we would have lost the whole thing. They were always trying to reduce risk as they perceived it and so take control away from the design team and give it to a contractor – but of course they [the big British contractors] ran a mile because it was using technology that is not common to them.'

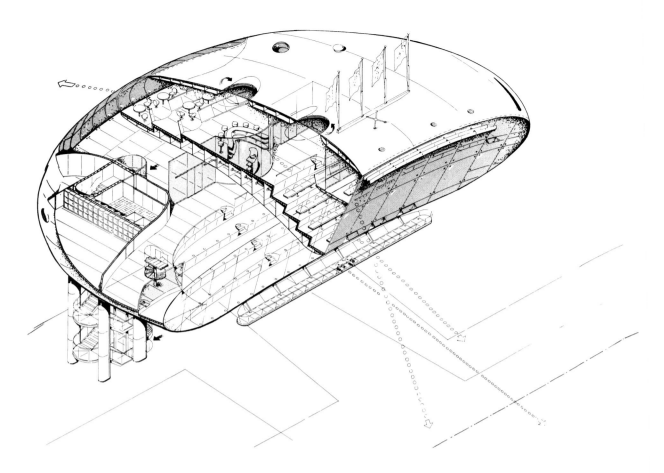

**Above:** The press centre's structure and accommodation.

**Below:** These sections were prefabricated in Europe and welded together on site.

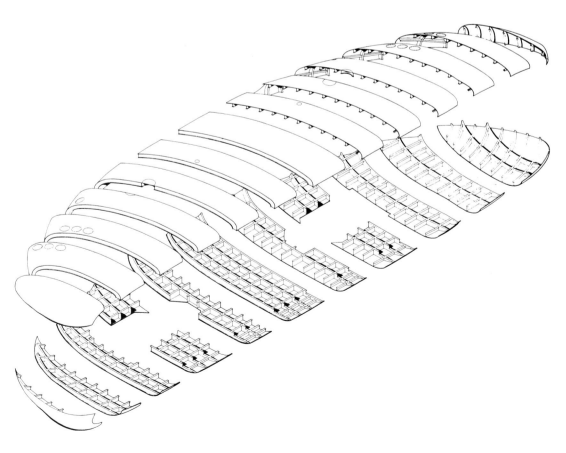

**Above:** The two legs are extensions of piles buried deep in the ground and carry the eccentric load of the centre above.

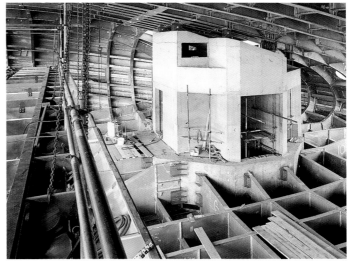

**Above left and right:** Although the external skin of the press centre suggests a monocoque structure, these views of the interior during construction show that it is actually an extension of traditional rib and skin boat building technology.

**Above:** The upper part of the centre temporarily stitch-welded together in the shop.

**Below left and right:** The skin during construction with a prefabricated section just moved into place.

**Above:** A prefabricated element in a sling about to be craned into position.

**Below right:** Computer model of the grille of ribbing.

**Below left:** A prefabricated section of the relatively flat underside of the press centre.

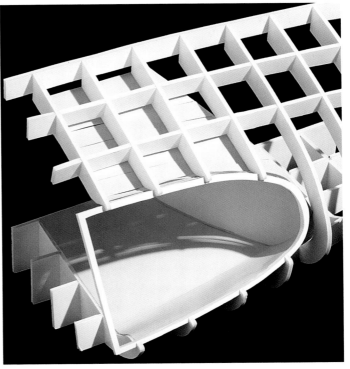

# Engineer: **Schlaich, Bergermann & Partner**

Spandau Station, Berlin, 1997

Architect: **von Gerkan, Marg & Partner**

**The new Spandau station is on the line between Berlin and Hanover. It is the first such project by the newly privatized German railway system, Deutsche Bahn AG. Its platforms are covered by four adjacent glass barrel vaults, strictly barrel shells, the longest of which is 430 metres (1410 feet) long. The vaults are unequal in length and span, their lengths reflecting the trains using the platform.**

The two middle vaults are wider, serving two tracks each, and the existing curve in the tracks is accommodated simply by making the almost straight vaults wide enough to cover the curve. The south vault is half the length of the others, and half the length of the adjacent vault has been omitted. As a result, what elsewhere is an arch construction becomes a curving cantilever along this section, partly propped by horizontal ties back to the adjoining vault. Running along the tops of each of the complete vaults is a strip of ribbed steel sheeting to facilitate ventilation.

In some respects the form of the station is a reworking of traditional big nineteenth century railway stations. The structure is a network of steel struts and diagonal bracing, fixed between big steel arches spaced 18 metres (59 feet) apart. The glass panes are a double lamination of 8 millimetre (1/3 inch) toughened, flat sheets of glass fixed to horizontal steel bars 60 millimetres (2$^1$/$_2$ inches) square. The arching assemblage of glass, horizontal bars, vertical struts and criss-crossing bracing cables is anchored every 18 metres (59 feet) to a big steel arch. Its function is less to actually support the gridnet than to break it into manageable sections, in much the same way that Matthew Wells at Techniker divided the Prague Castle glasshouse into three sections (see pages 124–33).

Spandau Station

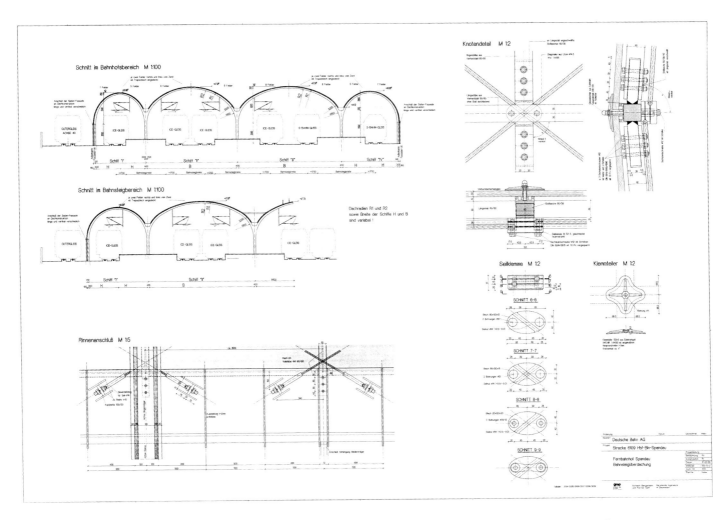

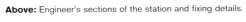

**Page 105:** At first sight this is a rather Victorian structure, until its delicacy and detail bring a better realization of how it works.

**Opposite:** An intermediate arch provides anchorage points for diagonal tension cables.

**Above:** Engineer's sections of the station and fixing details.

**Above left:** Cables are anchored at the intermediate arch while the main grid runs over the top.

**Above right:** Anchorage points for tension cables along the base of the vault.

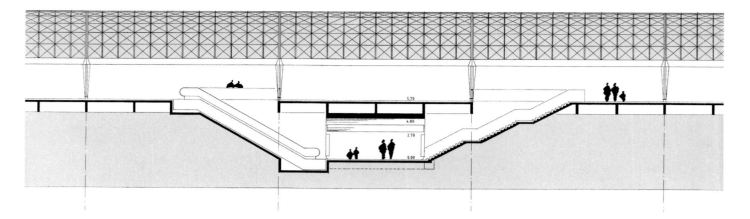

**Above:** Long section through the station with stairs leading down to the transverse concourse (opposite, top).

**Below:** Bracing cables are continuous across the structure and held in place with these simple friction clamps.

Spandau Station

**Above:** The main concourse is essentially a shopping street.

**Above left:** Aerial view of the station with platforms of different lengths related to the size of trains.

**Above right:** Perspective section along the underground concourse with the platforms and arches above.

# Engineer: **Neil Thomas of Atelier One**

Tower of Babel, Millennium Dome, Greenwich, London, 1999

Architect: **Mark Fisher**

**Mark Fisher, formerly in partnership with the engineer Jonathan Park, trained as an architect but has spent most of his career as a designer of fantastic sets for rock stars. These include all the Pink Floyd stadium shows and Rolling Stones tours during the last decade and a half. With the break-up of the partnership Fisher has collaborated with such engineers as Neil Thomas of Atelier One. Fisher specializes in interactive sets – all of which are taken down and re-erected many times during a tour.**

The set for Pink Floyd involved the building and destruction of an enormous wall across the end of the arenas in which the show was staged. In the Rolling Stones sets vast inflatable people unfolded on stage, while in the 1997 'Bridges to Babylon' show a self propelled 50 metre (164 foot) long bridge, designed by Neil Thomas, slowly cantilevered out from the main stage to connect with a secondary stage in the middle of the arena.

**Preceding page:** The Tower of Babel at full extension.

**Above:** As the main Babel tower rose, so did the middle tower with performers on safety lines on the spiral walkway preparing for the next stage.

Tower of Babel, Millennium Dome

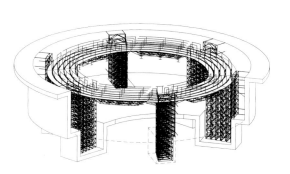

**Stage 1**

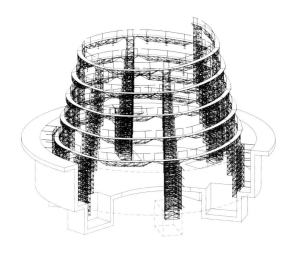

**Stage 2**

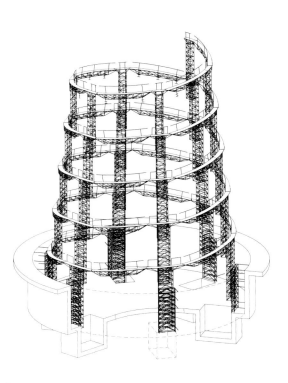

**Stage 3**

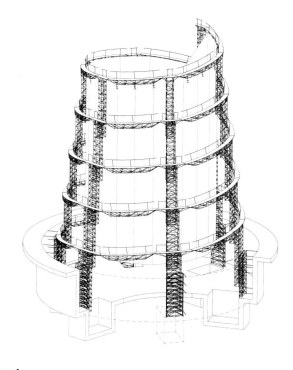

**Stage 4**

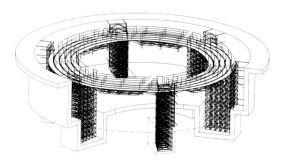

**Stage 5**

**Stage 1:** The collapsed tower with performers in place.

**Stage 2:** Half extension.

**Stage 3:** Full extension.

**Stage 4:** Drop sheets let loose.

**Stage 5:** Tower back in its pits.

**Above:** The tower begins to rise with crew members aboard.

The Tower of Babel was the centrepiece for a full scale performance by aerial dancers and gymnasts in London's new Millennium Dome. The show was devised by Fisher in collaboration with choreographer Micha Bergese and musician Peter Gabriel. The aerialists and a variety of Miró-like structures were suspended from a big ring truss hanging under the apex of the fabric dome (see page 41). At one stage during the show, and over a period of two and a half minutes, the Tower of Babel extended 24 metres (79 feet) upwards into the 50 metre (164 foot) high space from an opening in the apparently flat floor. Mark Fisher said, 'Neil and I worked out the look of the tower and designed it to look like a Victorian gas holder – and the traditional Babel symbol.' Before this a central lookout tower rose from the middle of the floor. Its many legs, at first splayed out, drew together as this watchtower emerged. When the encircling tower was fully erect big painted cloths were dropped from the top ring, enclosing the whole space. An additional engineering complication was that the tower also had a spiral walkway running round it which unfolded as it rose and from which performers abseiled.

The first engineering issue was whether it would be possible to dig sufficiently deeply into the foundations for the sections of the tower to be concealed under a movable lid in the centre of the site. Originally a marshland and a gas works, this was one of the most severely polluted sites in Europe. Engineers for the main dome had covered the site with a concrete cap and negotiations resulted in permission to go down only 5 metres (16 feet). So the five retracting segments of each of the six space-frame towerlets were lodged in six pits spaced around the perimeter of the big 5 metre (16 foot) deep, drum shaped excavation.

The towers were raised using a hydraulic primary drive system to get the retracted structure up beyond stage level, where performers climbed on board to attach rigging for the backcloths. Then a secondary hydraulic system was mounted in the first segment, which raised the remaining four using a system of drive chains looped over rollers at the head of the adjacent segment and attached to the bottom of the next one. As the second segment rose under hydraulic power the fixed length chains between the first and third, the second and fourth and the third and fifth all began to rise at a speed of around 33 millimetres (1 1/4 inches) per second.

**Above:** The tower descends into its pit.

**Below:** The five telescoping towerlets are each stored in individual pits. A pulley system raises the segments in turn.

In its retracted position the spiral walkway nests into a flat configuration sitting on top of the segments. It is attached to only three of these, the remaining three sloping down to the level below and attaching to their respective towerlets by rolling connections. This maintains the stiffness for at least half the towerlets while sustaining the illusion of a continuous spiral from ground to top.

Fisher comments, 'It was interesting because it was a cool piece of engineering which was watched by 6.5 million people and ran for 1000 shows without breaking down. The engineering team had to sort out the issues of wobbling several years before the Millennium Bridge and they deployed dynamic anti-resonance, anti-bouncing techniques. It was interesting because what we set out to do was dynamic and preposterous yet it not only looked architectural but was at a traditional architectural scale.'

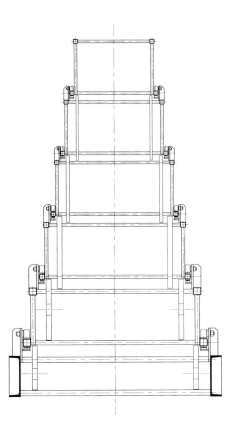

**Right:** Sequence of growth. The five mobile elements of each towerlet are stacked retracted in their pit. With the power on, the base cable pulls the bottom element up to its fixed position. The others, using their own interconnected system, then begin rising until each reaches its peak.

**Opposite page:** Structural fireworks.

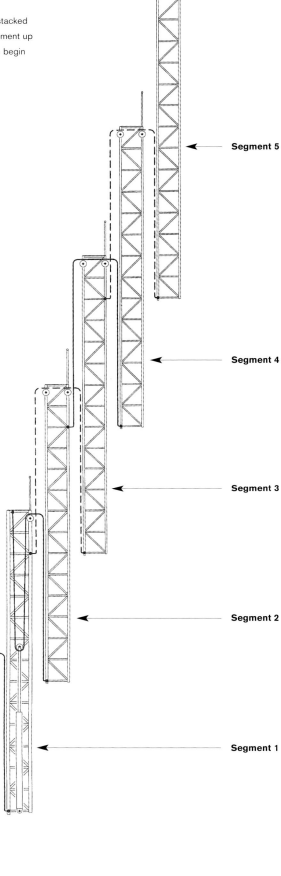

**Segment 5**

**Segment 4**

**Segment 3**

**Segment 2**

**Segment 1**

**Fixed base frame**

Tower of Babel, Millennium Dome

# Engineer: **Dewhurst McFarlane**

Pavilion, Broadfield House Glass Museum, Kingswinford, West Midlands, 1992

Architect: **Design Antenna**

**This glass museum is based in Broadfield House, a historic property guarded by the UK heritage policing authority, English Heritage. The authority insisted that any new building should allow the old one to be seen through it and so in some ways this all-glass structure is a nicely judged over-reaction. It formed part of a much wider refurbishment of the old building with galleries and a shop, a glass blowing studio and a sculpture court. At the time it seemed hardly possible to design such a glass structure.**

Extremely thick glass had been used for curtain walling as early as the 1980s – in Norman Foster's brilliant Willis Faber insurance building in Ipswich, for example, or the end wall of his Sainsbury Centre at the University of East Anglia near Norwich – but it was thought that glass was simply too fragile a material to be used for structural elements such as rafters. Engineer Tim McFarlane says that no one had really looked at the published data on the physical properties of glass and that even modestly sized glass performed quite well in shear and bending.

Pavilion, Broadfield House Glass Museum

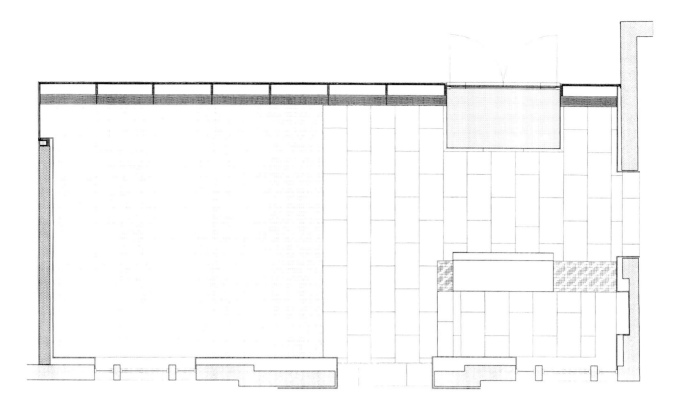

**Opposite above and preceding page:** The laminated glass beams connect with their laminated glass columns by interleaving the layers mortise-and-tenon fashion.

**Opposite below:** The all-glass box nestles comfortably into the wall of the old building.

**Above:** Plan with new free-standing flank wall on left, reception desk bottom right and walls of the old building bottom and right.

**Following pages:** Interior at night.

Although at the time it was believed to be the largest self-supporting, all-glass structure of its kind in the world, this is not a large building – only 11 metres (36 feet) long, 3.5 metres (11 feet) high and 5.5 metres (18 feet) wide – and it is attached at one side and the back to the outside walls of the old brick façade. Its structure consists of slender toughened glass columns, 200 millimetres (8 inches) deep and made up of a triple lamination, supporting 300 x 32 millimetre (12 x 1 1/3 inch) glass beams of similar configuration at 1.1 metre (3 1/4 foot) centres. To this basic structure are attached big, double glazed sheets of glass for the walls and triple glazed sheets for the roof. These also serve to laterally restrain the beams and columns. The upper 10 millimetre (1/2 inch) thick leaves of the roof panels oversail the external perimeter by 900 millimetres (3 feet). The separation between the two or three sheets of glass is achieved using high performance silicone sealants.

Glass inevitably presents environmental control problems, especially when the brief calls for the glass to be as clear as possible. The double and triple glazing goes some considerable way to minimizing energy loss and the panels have a silver soft coating which maintains the impression from outside that the glass is more or less clear. The underside of the roof glass is fritted with a pattern of ceramic dots, varied in some places to provide more shading. The parallel rows of fritting are echoed by a sandblasted pattern on the glass overhangs.

In a sense this is the primitive hut of glass architecture, the defining classic model of post and lintel structure in the purest and most immaterial of materials. Yet the glass is only 'almost' frameless. Where this is conventionally achieved with spider connections and clear silicone sealant, the use of multiple glazing layers almost always, as here, presents problems of restraining the various sheets in position – and of dealing with butt joints such as at the junctions between both columns and beams and their cladding panels. The visual end result here is of a structure of delicately framed glass panels.

Engineer Tim McFarlane has gone on to specialize in glass structures, including a sensational, bolted, cantilevering glass canopy over a Tokyo subway entrance by the SDG/Viñoly Tokyo International Forum building, and the Viñoly-conceived Kimmel Center for the Performing Arts in Philadelphia.

Pavilion, Broadfield House Glass Museum

# Engineer: **Matthew Wells of Techniker**

Orangery, Prague Castle, Prague, 1999

Architect: **Eva Jiricna Architects**

**This was the winning design by brilliant Czech expatriate architect Eva Jiricna with virtuoso structural engineer Matthew Wells in an international competition set by the new Czech Republic. The Orangery is a greenhouse for the Castle gardeners but is also used by the president, Vaclav Havel. He is an asthmatic who once worked on his plays in another greenhouse on the same site high above the city's pollution. For 15 years Wells had collaborated on all of Jiricna's High Tech projects including a series of enchantingly fragile steel and glass staircases.**

This glasshouse barrel vault too is fragile and glass and stainless steel but it is a big project, 100 metres (328 feet) long. It is divided into three separate climatic zones for wintering and growing different plants and giving Havel a place to work. The four cross frames (including the end walls) are trussed arches which act as stiffening rings against the tendency of thin shells to buckle inwards. This division into three also minimizes the stresses inherent in the vault.

The old conservatory leaned against an ancient wall which now was imbued with too much history and lacked the structural strength to form part of the new Orangery. Wells's solution was to devise a tetrahedral space-frame girder located just above and out from the old wall and running along most of its length. This girder is supported at the four cross-frame locations partly by diagonal props, which resist any tendency to buckling due to stresses in the glass vault. The rear length of the vault, strictly speaking a grid shell, is supported by this girder and at the front it rests on a steel-clad concrete edge beam. The cross frames have sliding joints because the extremes of temperature are between -20°C and +40°C and the structure moves as the temperature changes.

Grid shells are inherently low stress structures. In this one the waterproof layer is made up of big sheets of toughened laminated glass. They are frameless and are jointed and waterproofed with black silicone. They are flat, rather than curved, and there are eight facets running the length of the Orangery forming a 'circular' cross section. The bottom facet is heavily framed and contains glass louvers used in conjunction with butterfly vents at the apex of the vault to control the internal environment. At the corners of the glass sheets where four of them meet there is a spider fitting, integral with the nodes which connect each of the 48 millimetre (2 inch) diameter stainless steel tubes making up the grid shell. The gaps between the glass sheets are filled with silicone. Environmental controls include opening vents at the top and along the lower front of the vault; heating tubes located in the floor slab and along the front of the vault; and a set of silver automatic roller blinds which provide shade in the summer and some insulation in winter.

**Preceding page:** A simple grid shell but a wonderfully intricate structure of glass and steel.

**Below:** Computer model of the grid shell above an intermediate arch.

**Opposite:** The slightly awkward termination of the vault.

Orangery, Prague Castle

Matthew Wells of Techniker

**Above:** Computer model of the interior of the glasshouse with intermediate stiffening.

**Opposite:** Prague Cathedral through the glass slung under the external grid shell. The horizontal bands are retracted sunshades.

It was difficult to find the technical skills for this project in the fledgling Czech Republic so in the end a German company with a Czech subsidiary agreed to manufacture the components and run the project. The complicated structure, whose components had to be small enough to be handled without lifting devices, was put together by students using a big movable jig.

Jiricna and Wells spent, in his words, 'an inordinate amount of time' developing the standard node joint which not only connected four of the grid-shell struts but also the four sheets of glass hung below. Anyone familiar with the history of the space frame will know that the big design issue is always the design of the connecting node. Wells suggested a simple metal node with webs connecting the four fingers, while another design had the 'spider' as a four-legged stool rather than the frog pads finally adopted. At one stage Wells had suggested a pressed metal node for the spider supporting the glass but Jiricna was firm that these should always be in cast aluminium. Eventually they came back to more or less the original design.

Wells says that the main engineering problem with lightweight structures is to work out how they tend to buckle, especially under such loads as drifting snow. His engineers used a quite conservative dynamic analogy to highlight potential instabilities. The checking engineer in Prague used, says Wells, 'a much more sophisticated technique from classic theory than we did – cast into a home-made computer program. His analysis was beautiful and would have allowed a leaner structure.'

By this stage it was too late to re-design the structure. Wells had already suggested that by using differently sized members in response to changes in the stresses along the grid shell an interesting pattern might develop. Jiricna wanted an absolutely uniform structure so Wells's team, to satisfy their own sense of engineering propriety, slipped in different wall thicknesses to maintain the intellectual integrity of the structure.

This was not the only structural possibility discussed between engineer and architect. There were dozens. Early ideas included a grid of flat, glass-faced boxes with steel, nylon, glass or acrylic sides. These would be bolted together and forces would be taken up by cables running between the boxes. Another idea for a minimum structure was to deploy the glass as diagonal tension members, in a geodesic structure in which steel rods took the remaining tension forces and all the compressive forces.

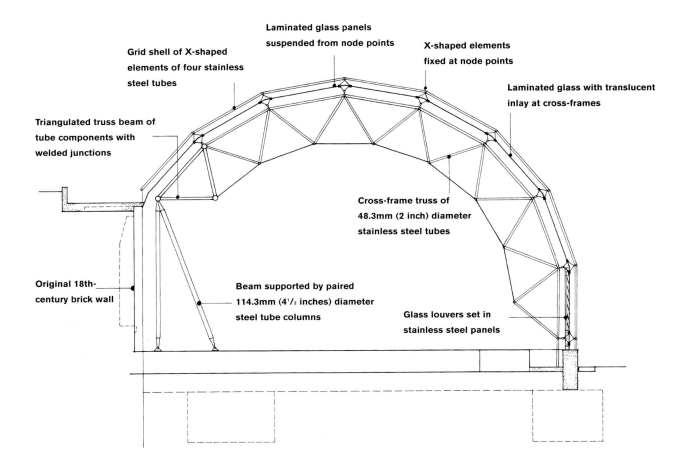

**Grid shell of X-shaped elements of four stainless steel tubes**

**Laminated glass panels suspended from node points**

**X-shaped elements fixed at node points**

**Laminated glass with translucent inlay at cross-frames**

**Triangulated truss beam of tube components with welded junctions**

**Cross-frame truss of 48.3mm (2 inch) diameter stainless steel tubes**

**Original 18th-century brick wall**

**Beam supported by paired 114.3mm (4¹/₂ inches) diameter steel tube columns**

**Glass louvers set in stainless steel panels**

**Above:** Section showing the old wall, left, the braces for the long triangular truss (first segment of curving lattice) and a standard lattice truss whose other end is supported on a concrete beam. The last panel has glass louvers set in stainless steel frames.

**Below:** One of the early ideas for the vault, using the old wall.

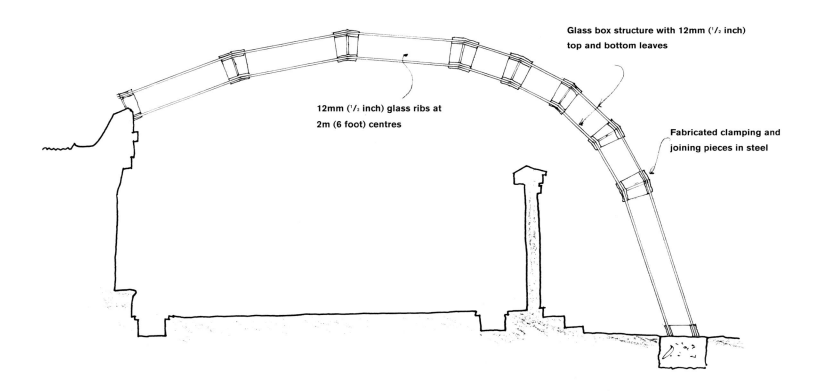

**Glass box structure with 12mm (¹/₂ inch) top and bottom leaves**

**12mm (¹/₂ inch) glass ribs at 2m (6 foot) centres**

**Fabricated clamping and joining pieces in steel**

**Above left and right:** Two views of the spiders which are part of the standard grid-frame node and join the corners of four adjacent panes of glass.

**Below:** Like a better mousetrap, the node which joins elements of the grid shell looks very simple though it is the outcome of a lengthy design process. The spider which fixes the glass panels is attached below it.

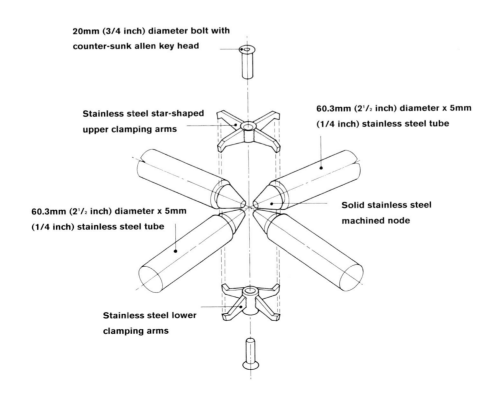

**20mm (3/4 inch) diameter bolt with counter-sunk allen key head**

**Stainless steel star-shaped upper clamping arms**

**60.3mm (2¹/₂ inch) diameter x 5mm (1/4 inch) stainless steel tube**

**60.3mm (2¹/₂ inch) diameter x 5mm (1/4 inch) stainless steel tube**

**Solid stainless steel machined node**

**Stainless steel lower clamping arms**

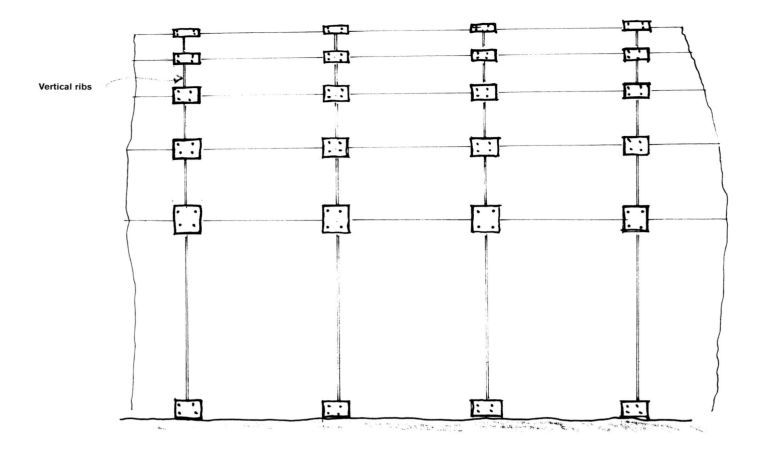

Vertical ribs

**Above:** An early all-glass proposition by Techniker. It is made up of a series of glass boxes clamped together at the corners to form the general structure of the vault opposite top left.

**Below:** Details of clamping (left) and module construction (right).

**Top and bottom cover plates –
screwed to clamping piece top plate**

200

**Chamfer glass corners
at junction to avoid bolting
through glass**

**Module construction: single or laminated
sheets provided top and bottom to
perimeter diaphragm**

Orangery, Prague Castle

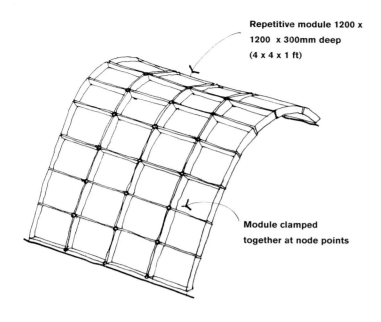

Repetitive module 1200 x 1200 x 300mm deep (4 x 4 x 1 ft)

Module clamped together at node points

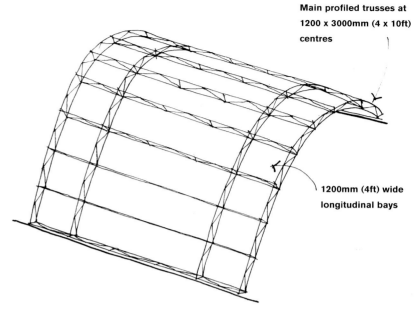

Main profiled trusses at 1200 x 3000mm (4 x 10ft) centres

1200mm (4ft) wide longitudinal bays

**Above, left and right, and below:** Some of the other ideas explored by Techniker and Jiricna: an alternative to the glass boxes (above left) using a braced metal framing (above right), and a series of possible rib formations (below).

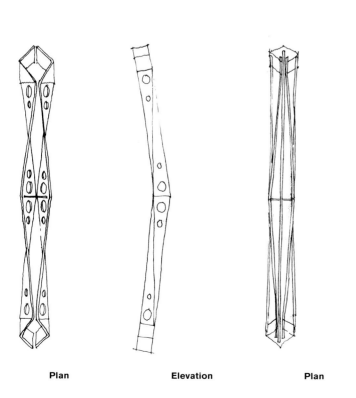

**Plan**

**Elevation**

**Plan**

Rods

**Elevation**

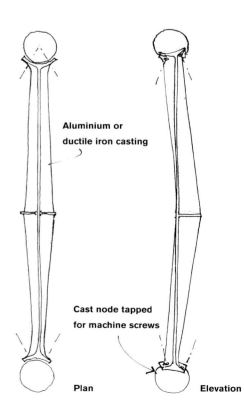

Aluminium or ductile iron casting

Cast node tapped for machine screws

**Plan**

**Elevation**

# Engineer: **Santiago Calatrava**

## City of Science, Valencia, 2000

**These two buildings set in a vast waterscape are the first of a group that the Spanish engineer-architect Santiago Calatrava has designed for the new City of Science district in his home city of Valencia. The first to be completed was the planetarium and the second the intricately skeletal, 241 by 104 metre (790 by 341 foot) science museum building; an arboretum is about to start and an opera house is due for completion in 2003.**

The planetarium is a quasi-open hemisphere with a pair of big triangular openings either side on the axis of the broad walkway. The other sides, facing the water, are glazed. Their lower section is an all-glazed version of Calatrava's famous opening warehouse doors at the 1985 Ernsting warehouse in Coesfeld, Germany, whose geometry is suggestive of an opening eyelid – here on a vast scale. The complicated and brilliant geometry involves an upper and lower section which are hinged along their curved junction and fold upwards.

  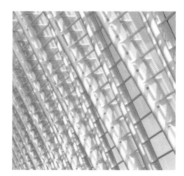 

City of Science

This is a Calatrava favourite which has been used elsewhere, for example at the base of his communications tower in Barcelona, and was licensed for a building in the UK. It is the signature animated image on his office website where, as here, the lids fold up to reveal an eyeball. At Valencia the eyeball is actually the hemisphere of the planetarium, which is enclosed by the big quasi hemisphere. When the articulated glass 'eyelid' opens on either side, the inner hemisphere is reflected in the water and appears to become a complete sphere.

The science museum structure alongside in the big lake has a very complex but constant cross section with glazing on the north side and aluminium cladding on the south. The main wall of curved glazing makes a very clear reference to the ridge and furrow glazing system of Joseph Paxton's 1851 Crystal Palace – a purer form of which was developed by engineer Tim McFarlane for Rafael Viñoly's almost contemporaneous Kimmel Center for the Performing Arts in Philadelphia. At Valencia Calatrava has devised a more complicated pattern in which the troughs and ridges taper alternately. This produces relatively low profiles at ground level and high, haunched curves at the top, where the glass forms fit snugly below the sharp peaks along the upper eaves of the curving roof. On the south side these peaks are repeated, but at right angles to the eaves, forming an extraordinary, repetitive, skeletal ensemble whose purpose is not always easy to grasp. Inside, the great glazed north wall has the quality of the side aisle of an imaginary gothic cathedral, with the vast main piers branching up into the intricate concrete tracery of the roof and supporting several floors of exhibition space.

**Preceding page:** Calatrava's Gaudiesque exuberance which worries right-thinking engineers and architects.

**Right:** The great north-facing glass wall of the science museum has a ridge and furrow profile but at a scale its early nineteenth-century inventors could hardly have envisaged.

**Above top:** Even during construction Calatrava's forms have extraordinary fluidity.

**Above centre:** The organic support structures which look like tree trunks and branches express the collection of loads from various parts of the roof down to the ground.

**Above:** The curving shell of the planetarium enclosure, with the eye just beginning to open on the left.

City of Science

**Above left:** The ridges and furrows with plainly Gaudiesque terminations.

**Above right:** The planetarium by night with Calatrava's famous opening eyelid in the closed position.

**Above:** The buildings are set in a vast plane of water.

# Engineer: **Ingenieursbureau Zonneveld**

Waterland, Burgh-Haamstede, 1997

## Architects: **Lars Spuybroek (NOXArchitekten) and Oosterhuis.nl**

**This design is by avant garde architect Lars Spuybroek of NOX with fellow Rotterdam architects Oosterhuis.nl and J. P. Van der Windt of engineers Zonneveld. It is a celebration of water, providing practical information as well as a series of pleasurable water-related assaults on the sensations. The result is an overwhelming experience of image and atmosphere. Inside the windowless pavilions nothing is orthogonal or horizontal, even the floors. Walls and floors are ambiguous and merge into each other, while the lighting and air handling ducts snake across the ceiling, scarcely noticed in the cobalt blue artificial light. There are mist sprays, a frozen glacier tunnel, a pool draining across the floor, a stroboscope-lit rain basin and a well.**

The two-part building is located on a former construction island, Neeltje Jans, which is part of the Oosterschelde dam in Zeeland, the south-west province of the Netherlands. One pavilion demonstrates salt water, the other fresh. Kas Oosterhuis designed the black mass of the saltwater pavilion which is connected to Spuybroek's giant silver 'slug' and hangs right over the man-made gravel beach.

   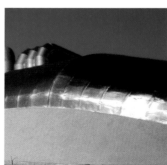

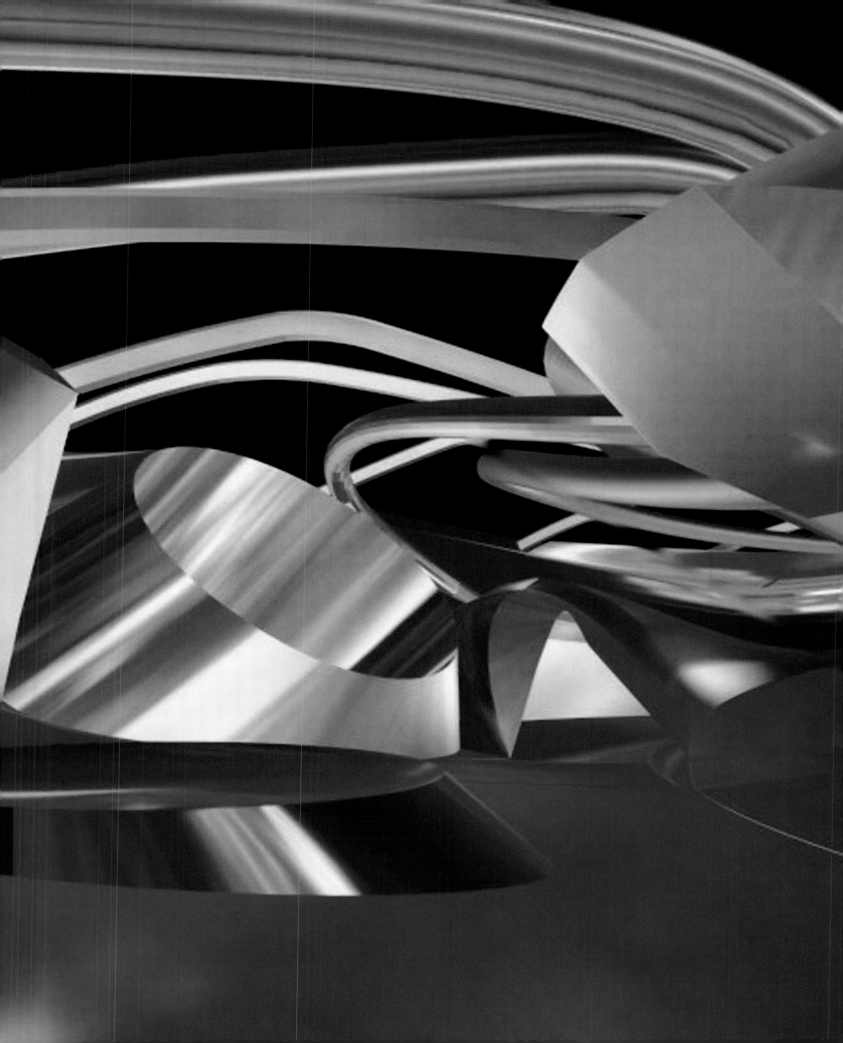

The structure of this extraordinary group of convoluted shapes is curiously conventional. As Van der Windt, the engineer, points out, it has been realized with relatively traditional building materials. 'At first the architect concentrated mostly on designing the geometry. The initial phase of the design process took place with the computer; in later phases, scale models were also used. The subsequent assignment for the structural engineer was not so much working on the forces in play, but rather finding the right materials that could follow the curved shapes.'

It was not a particularly straightforward process. As the engineers argue: 'If one thing has become apparent, it would be that the present environment is hardly ready for this 'liquid' architecture. The design was not the problem; the limited possibilities of the materials posed the biggest problem. Initial solutions seemed elegant, but turned out to be too expensive, to pose risks or to take up too much research time. Another handicap was that the communication between designers and suppliers passed through the building contractor. Suppliers are usually more inclined than contractors to use experimental materials. All things considered, more traditional materials and structures must be used, greatly contrasting with the architecture. Instead of creating shapes, it deforms materials. There is a world of difference between dream and deed.'

The shape of the original design for the saltwater pavilion was a smooth, flowing amoeba shape. The imagery the design team used was that of a stranded whale. The built form, designed using CAD, has turned the curves into hard chines or edges and has been raised 1 metre (3 feet) above sea level. The structure is made up of straight I-section arches with profiled steel decking spanning between them, coated with a sprayed insulation and then a black abrasion-resistant polyurethane coating. The internal wall surface is stretched canvas duck.

The design team had hoped that the structure would be entirely concrete or constructed by a steel or shipbuilding company. Asked to tender, both kinds of firm asked for greatly inflated prices because the project was unusual. The all-concrete structure turned out to be too expensive as well. Eventually the arched beam solution was adopted, the beams being connected longitudinally with horizontal circular steel tubes. The saltwater pavilion, which hangs approximately 12 metres (39 feet) over the surface of the water, is supported by cantilevering the concrete floor out and adding steel beams to stiffen the structure. Back on land the tendency of the structure to fall into the sea is resisted by thickening the slab very considerably and tying it to a group of pile heads.

Waterland

Lars Spuybroek designed the freshwater 'slug' on computer, generating and manipulating a series of sections. Each had different curves and each fitted into the sinusoidal plan. That turned out to be also a diagram for the final structural solution, with a series of arches averaging 16 metres (52 feet) across defining the cross-sectional shape at any point. Once the form had been established, it was frozen while the search went on for an appropriate covering. After much research and debate the design team came up with a thin, flexible stainless steel bonded to a bituminous material. It was used in the petrochemical industry as a covering for ducts and a lining for dangerous substances. It came in relatively narrow rolls and the substrate of adjacent pieces could be welded to each other, which made it especially suitable for the complex three dimensional shapes involved.

The arches are connected together by C-section cold rolled purlins, to which the three layers of plywood substrate are screwed ready for fixing the final flexible covering, the stainless steel/bitumen compound. Some of the arching I-section beams (the ends of the C-section purlins bear on the inner base flanges either side) have very tight radius curves. In order to achieve these, the relevant beams were cut down their middles, curved independently between them and welded together again. The purlins have to be at right angles to the ply substrate. Some are therefore anything but perpendicular so the twisting stresses thus created had individually to be recalculated and the purlins made larger.

Setting out the structure was difficult for the engineers because the prefabricated arches had to fit exactly on the concrete foundation strips. It was a problem of not only horizontal location but vertical position as well. So after the concrete pour and while the finishing off work was being done, the exact level was indicated by the architect on the spot. This working method, so the engineers point out, proved much easier than communicating via drawings.

**Page 141:** Computer rendering of the freshwater pavilion interior.

**Below and opposite:** The exterior of the shiny slug (the inverted 'boat' of the saltwater pavilion is attached to the right): an enigmatic form which nevertheless hints that something extraordinary may lurk inside.

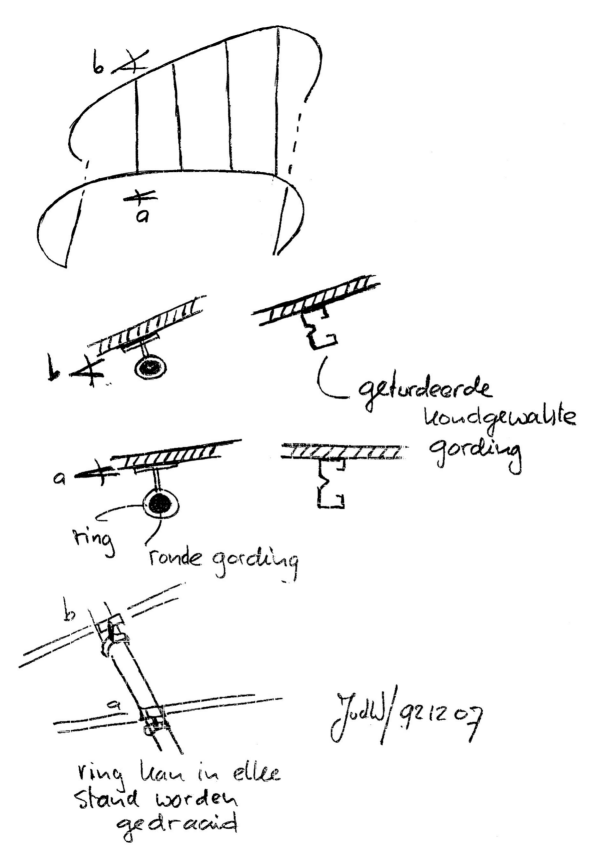

geturdeerde
koudgewahte
gording

ring

ronde gording

JvdW/ 92 12 07

ring kan in elke
stand worden
gedraaid

**Above:** Engineer J.P. Van der Windt's early thoughts about the effect of eccentric loading around the attachment of the skin to the basic frame.

**Above:** The dream-like interior of the slug with changing floor levels and textures.

**Below, left and right:** Computer studies of the slug and its structure.

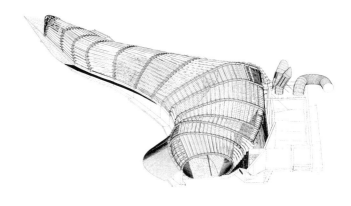

Waterland

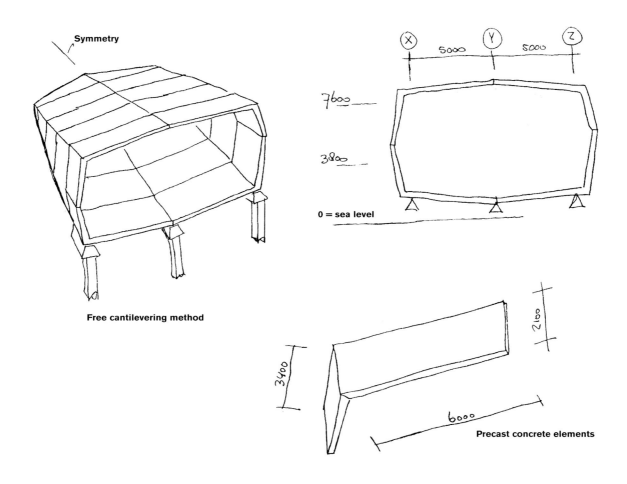

**Symmetry**

**Free cantilevering method**

0 = sea level

5000  5000

7600

3800

3400

2100

6000

**Precast concrete elements**

**Above:** The structure of the saltwater pavilion.

**Opposite, top and bottom:** The saltwater pavilion overhangs the artificial shore like a beached, inverted chine-hull boat.

**Below:** A computer image delineates the fantastic fluidity of the internal spaces.

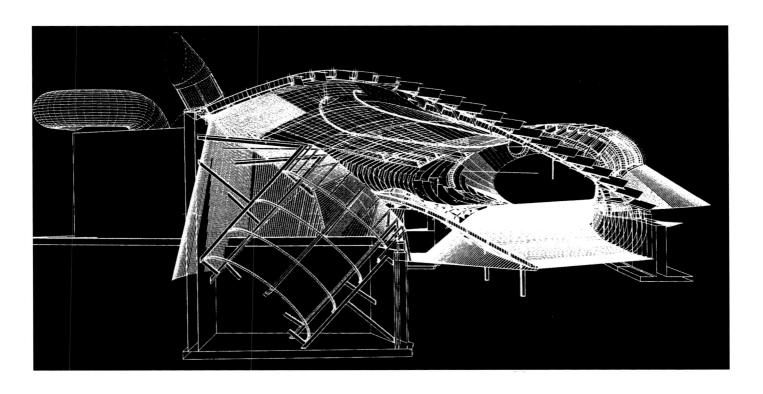

# Engineer: **Skidmore Owings & Merrill (Chicago office)**

## Guggenheim Museum, Bilbao, 1997

## Architect: **Frank O. Gehry & Associates**

**This museum, together with other significant engineering-architecture by practices such as Santiago Calatrava and Foster Associates, has transformed this once decaying Spanish industrial port. Few walls are flat or vertical. The configuration of its plan and internal levels and its swooping, free, titanium-clad forms were difficult for the engineers to describe and analyze because none of them belonged to any familiar geometrical system.**

The sensational building is in fact a merged cluster of 11 building forms arranged around a 50 metre (164 foot) high atrium. The engineers have described it thus: 'The architectural theme of fractured and irregular building masses was explicitly at odds with the normal structural engineering precepts of stability, organization and regularity in order to achieve a design which is efficient and cost effective.' It needs to be said that one engineering commentator, Henry Petroski, pointed out that the Statue of Liberty is an analogous case: the support structure for its convoluted skin design called for the engineering genius of no less than Gustave Eiffel.

Guggenheim Museum, Bilbao

The original thought was that the basic structure would be concrete. But the engineers, says SOM's John Zils, had a feeling that steel would be better because of the expense and near-impossibility of casting concrete in multi-formed moulds more than 30 metres (98 feet) in the air. The existence of a well established ship building tradition in Bilbao was also relevant. After much geometrical analysis the engineers came to the conclusion that it would be possible to support the complex skin curvatures using a close-spaced 3 metre (10 foot) square grid. This was dense enough to roughly conform with the curved surfaces and a good dimension for transporting elements from the fabrication shop to the site by road.

Each of the steel members making up the basic grid was different. The vertical members were I-section joists, one flange facing outwards, the horizontal members were square tube and the diagonal members circular tube. At first it was assumed that joints would be welded but the decision was taken to bolt them instead – not least because with computer cutting of steel members there was much less chance of bolt holes not lining up. Unlike most steel structures, which have bracing only where it is strictly necessary, every grid square has a stiffening diagonal member. This, plus the raking, twisting form, provided the structure with sufficient stiffness as it was being erected for there to be little need for supporting scaffolding. Although the members are of different lengths, from grid to grid the assembly and layout remain constant – considerably facilitating calculation and CAD-CAM manufacture.

One critical element was the design of the nodal joints. It had to allow adjacent members to have angle changes to fit the curves of the structure. The rule of economy of scale/repetition dictated that the joint should be as universal as possible. SOM's design is based on horizontal bearing plates, shop welded to the vertical I columns to take the bolted horizontal members, with gusset plates welded on to take the bolts for the diagonal circular tube struts. This fitted in with the fabrication programme of

creating 3 metre (10 foot) high truss 'bands' – essentially grid-high sections of the structure which turned out to have a high level of stiffness – which could be transported to the site where they would be bolted to the adjacent structure. To ensure accurate alignment the 'bands' were trial-fitted in the workshop against adjacent assemblies.

Although stainless steel had been the first choice for the cladding, the cost of titanium suddenly dropped just as the bids were going out. The design team took the opportunity and used 0.38 millimetre (1/64 inch) titanium sheet, choosing it for its strength and durability and, one suspects, for its apparently exotic character. It had been decided that the skin should always be 300 millimetres (1 foot) away from one of the 6000 nodal joints (which was how their co-ordinates were established in the first place). The skin is carried on a lightweight secondary galvanized steel frame of open channel sections. These are fixed to horizontal tubes mounted on adjustable outriggers which are attached to the horizontal square tubes of the main structure. They define the curvature of the surface. The thin titanium sheets are fixed with stainless steel screws over a self-healing membrane coating a 2 millimetre (1/10 inch) galvanized steel substrate. Inside, the wall surfaces are mostly plaster, the 1 metre (3 foot) thick zone of the wall being used for insulation and as a continuous duct space for services.

In designing the structure the original Gehry model was digitally scanned to produce a three dimensional wire frame model. A French computer program, CATIA, used to design Mirage fighters and the Boeing 777, enabled Gehry to make final adjustments to the whole form of the building, then carry out structural analysis in Chicago and cut the steel frame members in Spain. For a variety of reasons the model could not be used to cut and form the titanium sheet but it was used to track and establish the location of panels. The same program was used to cut double curvature surfaces in the considerable amount of limestone cladding. Working in this unbroken electronic chain from architect's model to fabrication shop meant that everyone operated from the same data.

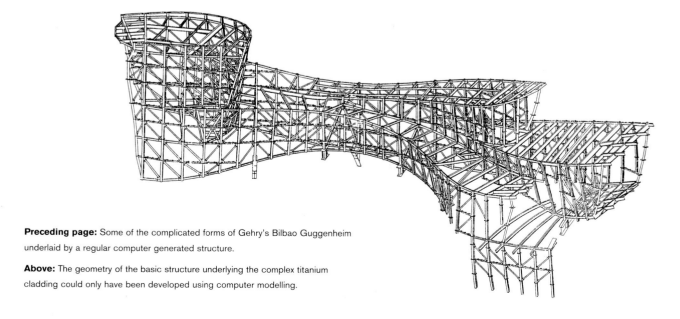

**Preceding page:** Some of the complicated forms of Gehry's Bilbao Guggenheim underlaid by a regular computer generated structure.

**Above:** The geometry of the basic structure underlying the complex titanium cladding could only have been developed using computer modelling.

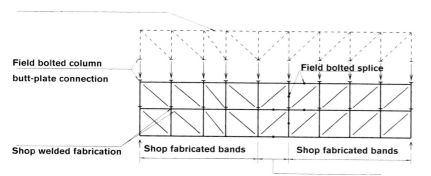

Field bolted column
butt-plate connection

Field bolted splice

Shop welded fabrication

Shop fabricated bands

Shop fabricated bands

Field bolt on the ground
Shorter bands together

**Above left:** The basic structural module, called a truss band, is shop fabricated and stacked on site. Connections are only at column butt-plates.

**Above right:** Fixing the titanium panels.

**Above:** The basic steel structure.

**Above:** The vertical bars supporting the titanium panels are fixed to horizontal tubes offset to the correct position from the basic structure.

**Above and below:** The complicated forms of the building are supported by an equally complicated rectilinear steel structure with appropriate curved steel elements for special volumes.

**Opposite top:** The inventive finished spaces created from the structural maze (see below left).

**Opposite bottom:** One of the tower interiors (see below).

Guggenheim Museum, Bilbao

Skidmore Owings & Merrill

**Above, left and right:** The smooth, curving plaster internal walls and external titanium cladding appear effortless. They give no hint of the complex geometry underneath, made up of individually cut steel members.

**Below:** Despite its beauty this is a functional computer model of the cladding. The model was eventually used to cut the individual titanium sheets.

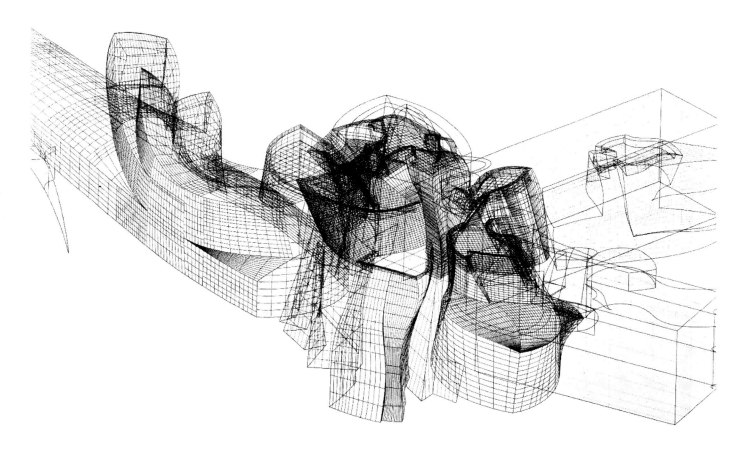

Guggenheim Museum, Bilbao

**Above:** The museum is a building with an extremely complicated surface and form.
Here it slides under an adjacent bridge and terminates with a great split wedge.

Skidmore Owings & Merrill

# Engineer: **M.G. McLaren and Buro Happold**

## Carlos Moseley Music Pavilion, New York, 1991
## Architect: **FTL**

This temporary fabric structure was designed by a team led by FTL's Nicholas Goldsmith, with engineers M.G. McLaren and Buro Happold and the acousticians Jaffé Acoustics. It was intended for summer music performances by the Metropolitan Opera and New York Philharmonic in New York parks. Its primary function was to provide a sheltered stage area for musicians and performers. The unusual twist was that it had to be capable of being shipped in and erected in a few hours – and taken away at the end of a performance.

The brief has a long history in the tradition of fairgrounds and circuses and, more recently, in arena rock shows – though with these the basic structural supports are normally in place. Here the problem was how to put up a structure in public parks and leave no trace afterwards. The general propositions of moving and transient architecture had been discussed by the British Archigram group and Cedric Price in the 1960s (both fathers of High Tech architecture) and this project is a real life manifestation of the earlier theory.

The trailers arrive on site with two folded front trusses which are positioned around the stage. The tractors are driven away and trusses cranked up.

The trusses crank upwards and begin to unfold.

The unfolded trusses, now locked into the form of single struts, are laid diagonally across the stage and connected to the rear truss which is still folded.

The rear truss begins to unfold and raise the whole structure.

The tripod half erected.

The crew begin to raise the fabric roof.

**Above:** The stage and canopy ready for a performance. The three skeletal structures to the right are units of the radio controlled sound system.

The whole structure is brought into a park on seven trucks which also provide a base for the structure. Three of them carry the three trusses which form the 20 metre (68 foot) high tripod supporting the fabric roof. These are hinged at mid point and positioned at predetermined angles at two corners of the 24 x 12 metre (78 x 40 foot) stage. This was earlier unpacked on to a simple steel structure resting on big pads. With the tractor units driven off and the trailers jacked up on big temporary foundation pads, the two front trusses are laid out flat across the stage and locked rigid. They are attached at their tops to the top of the third truss which is still folded and lies at right angles across the middle of the stage. Its tractor, still attached to the trailer, is driven away from the stage and as it does so it unfolds and gradually lifts the other two trusses up. The unfolding is achieved by a system of paired hydraulic rams forming an integral element perpendicular to the middle of the truss at its hinge. After it is locked, the now-rigid tripod can support the fabric roof and the lighting rig beneath it. The main support points are near the apex of the pyramid and the ground anchoring points are located on the trailers – now augmented by the two trailers which carried the stage positioned at the other two corners.

The taut, semi-translucent fabric roof has been shaped to act as a sound reflector and a screen for night time projections – when the trussed tripod effectively disappears. Sound is controlled by a specially designed system of radio transmissions to audience-based speakers.

The structure has had a long life and the system was redeployed by FTL in a 1995 relocatable structure used at the Atlanta Olympic Games. Here a big tent was suspended from a trussed pyramid rather than a roof.

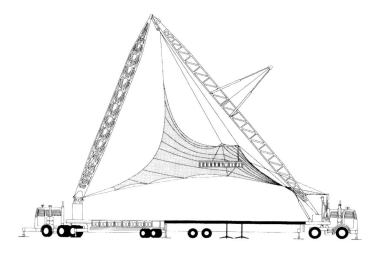

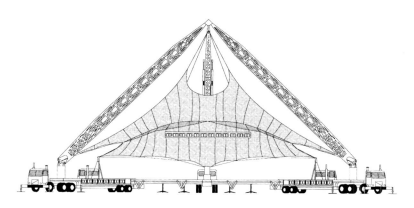

**Above left:** Side view of the pyramid showing the hydraulic assemblage at mid point on the rear truss with the roof and lighting rig slung underneath.

**Above right:** Front view of the pyramid and roof.

**Below:** Plan of the pavilion with all the anchorage trailers in place showing attachment points near the bases of the three trusses and two additional points at the rear on the trucks which carried the stage structure. In practice the tractor units were usually driven off and the trailers on which the base joints for the trusses are fixed were supported by big flat temporary foundation pads.

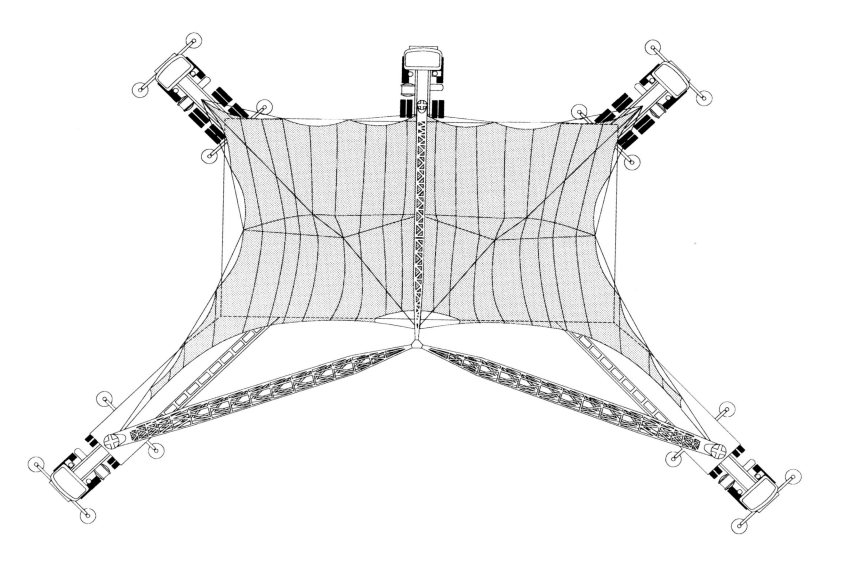

Carlos Moseley Music Pavilion

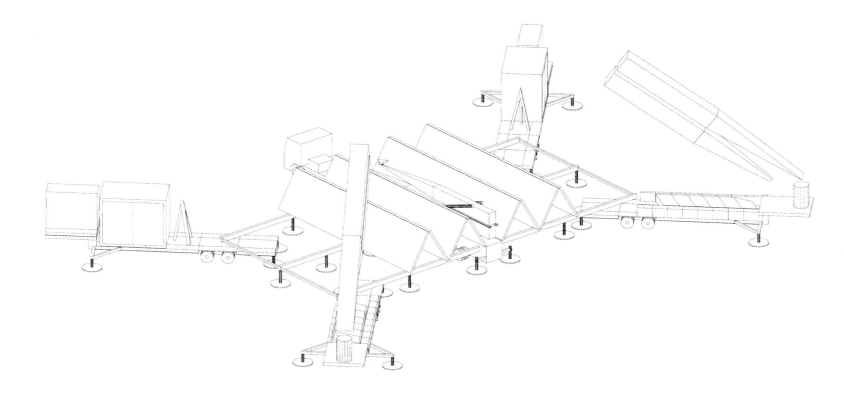

**Above:** The deck comes on two articulated trucks, one with the steelwork for the two-part base frame and temporary foundation pads, the other with the stage floor in two concertina-like sections. The latter drives in between the two frames and the floors are unfolded either side.

**Below:** The same slot is used by the trailer carrying the power-articulated rear truss. As it moves out the upper section of the truss begins to rise.

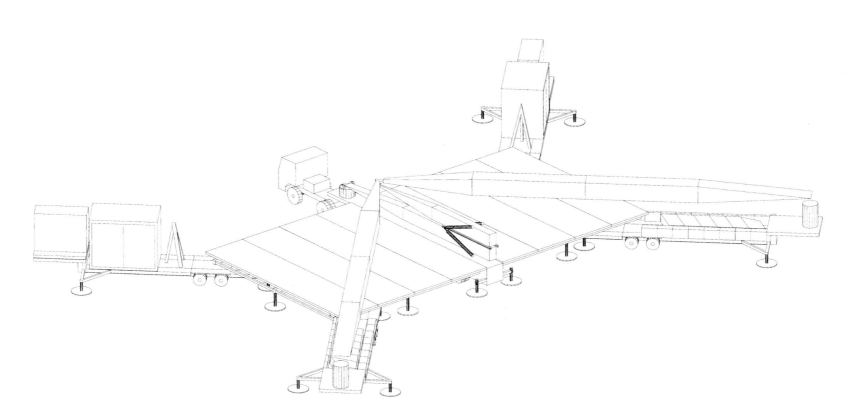

# Engineer: **Jane Wernick of Ove Arup, Allot & Lomax, Atelier One and Infragroep**

London Eye, London, 2000

Architect: **Marks Barfield**

**The London Eye, located on the south bank of the Thames opposite the Houses of Parliament, is the world's biggest observation wheel, put up to commemorate the new millennium. It was conceived by the architectural practice Marks Barfield in a competition run in the early 1990s by a British newspaper and an architectural pressure group. Nobody won the competition but the architects pressed on with their proposal and eventually saw their design in operation in the early months of 2000.**

Unlike most Ferris wheels, which have supports on either side of the axle, the London Eye wheel has spokes like a bicycle and is supported on one side only. The wheel is 135 metres (443 feet) in diameter and the whole load is carried by a giant bearing which cantilevers from a pair of inclined props leaning out over the river and is restrained by four massive cables anchored in the ground behind. The 32 capsules, each comfortably holding two dozen people, are glazed, egg shaped structures fixed to the outer side of the wheel. The wheel is rotated by a drive unit seated at its base.

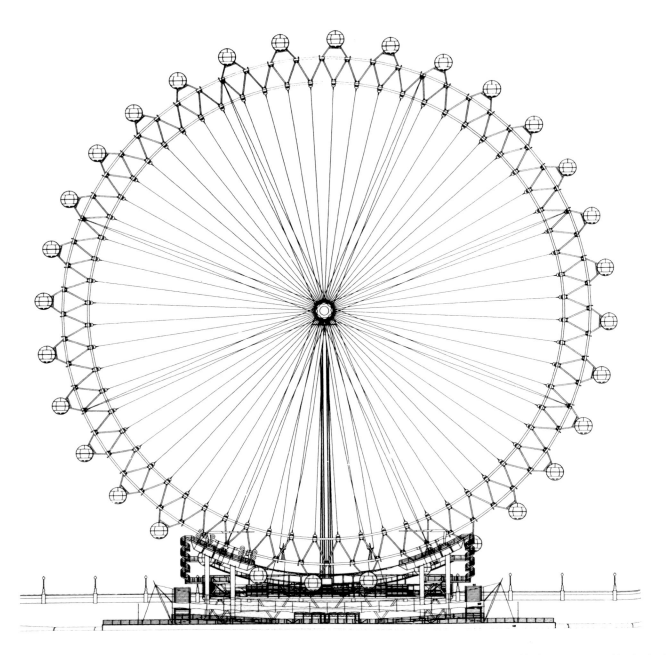

**Preceding page:** The Ferris wheel transformed structurally and conceptually into an instant London monument.

**Above:** The great wheel's spokes are arranged in the same way as a bicycle wheel. Its rotation is powered and regulated by friction rollers at the bottom.

The concept design was developed with an old friend, Jane Wernick, who was an engineer with the world's largest consulting engineer, Ove Arup & Partners. Wernick and colleagues at Arup worked on infrastructure, environmental impact and some early capsule ideas for the concept design. The general structure of the wheel had been worked out but no stress analysis had been done. The architects had borrowed against their house to work on developing the design and gaining planning approval, which was finally agreed at government level. Eventually British Airways agreed to back the scheme and took a half share in the enabling company, the London Eye Company.

British Airways wanted a turnkey operation and negotiations began with construction companies big enough to handle the project. Most wanted to revert to standard Ferris wheel construction techniques and eventually the last constructor begged off. In the end the contract was run by a construction manager, with the great Dutch steelwork company Hollandia taking responsibility for the engineering and fabrication of the wheel.

The French company Poma, which had an excellent safety record, was prepared to do the design work, prototyping and production of the capsules, which by this time had been reduced in number from the original 60 to 40 to the final 32. Julia Barfield and David Marks found

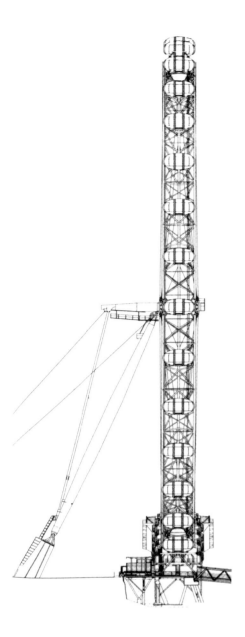

**Above left:** The wheel is hung from a great axle supported on two canted props tied back into ground anchors.

**Above right:** The cables have a strand omitted to provide a path for potentially hazardous water drops to run off.

that the capsules could accommodate far more people than traditional space standards seemed to suggest. Engineers for Poma were the Infragroep.

With the two major contracts established, final engineering design and stress analysis could begin with engineers Allot & Lomax. The engineering issues were complicated because there was no precedent for a structure of this kind and size, so conventional codes were not of much help. Allot & Lomax had to deploy considerable lateral thinking and advanced computer programs in evolving their final engineering proposition. Because the structure was in engineering terms a new one and

passenger safety was an overriding issue, there was particular need to check everything. As the engineers said, 'There must always be an underlying concern with any novel structure that something has perhaps been overlooked. Rather unusually…it has been possible to develop a more wide-ranging strategy demonstrating structural integrity… From the beginning, this entire process was collaborative and transparent with all parties contributing.' This process also involved a great deal of testing, and intensive monitoring has been built into the working wheel.

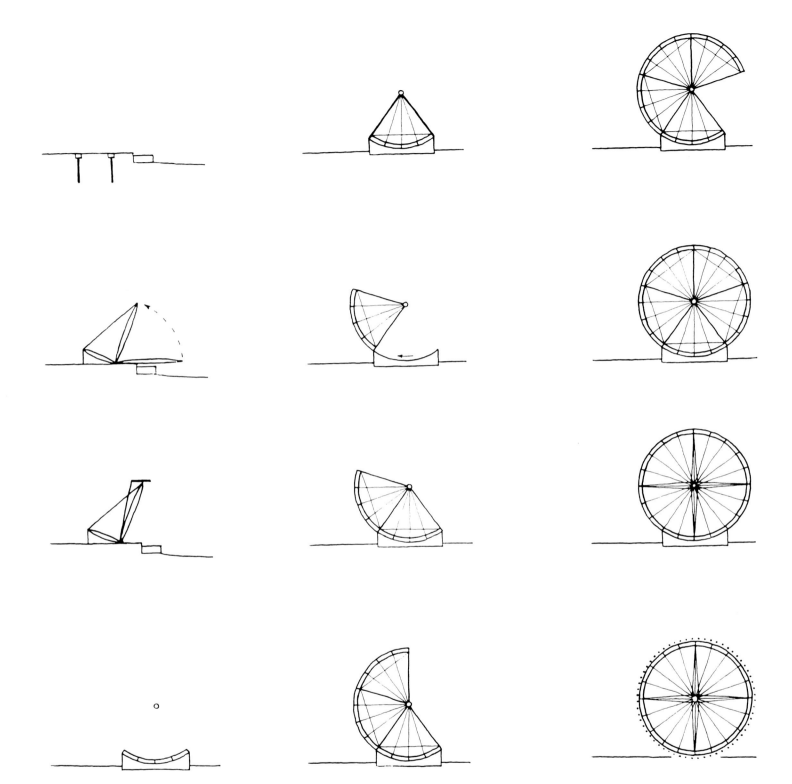

**Above:** Early drawings by Arup engineer Jane Wernick showing a suggested sequence of erection, with another series establishing the basis of the design of the wheel and its suspension.

**Opposite:** The structure was assembled flat and occupied a third of the width of the Thames.

London Eye

Jane Wernick of Ove Arup, Allot & Lomax, Atelier One and Infragroep    167

London Eye

In the final design the rim is a triangular open lattice truss in the form of a circle. This reflected the architects' concept and was a structurally efficient solution, providing adequate strength, torsional resistance and potential for load redistribution. The rim is attached to the hub by radial spokes which, as in a bicycle wheel, are arranged and pre-stressed so all 64 cables are under some kind of tension whatever the rotation of the rim. This puts the rim in uniform compression – modified by gravity so rim compression is actually lowest at the top of the rim. Since the pre-stress load on the cables was potentially very significant the engineers selected a load factor as low as possible, so in practice only a few cables slightly slacken as the wheel rotates.

The engineers had noted the effect of retained rainwater and ice on cable stayed structures, which effectively changes the shape of the cables and allows vibration. The cables' natural frequency and inherent damping also affect this vibration. Here dampers have been added to each cable and one strand of wire omitted to create a small spiral path down the cable to discourage water droplet alignment.

Wind loads were significant, not least because of the possibility of them destructively amplifying the natural frequency of the structure. The solution was to dampen the wheel structure by attaching 64 tuned mass dampers (TMDs). These are fixed across the two outer rings of the rim truss where they appear to the casual observer to be part of the structure. They are hollow tubes with low friction inner walls containing a spring attached to a mass fitted with nylon wheels to make movement easy and minimize noise. The engineers fine tuned the TMDs when the wheel had been erected.

The prefabricated elements of the wheel were constructed in Hollandia's waterside shops in Rotterdam and then brought by barge across the North Sea and up the Thames, where they were assembled into a horizontal wheel on temporary supports in the river bed. After some well publicized false starts, the wheel with its hub and A-frame support legs was finally raised into position. The remaining work on the drive system and the attachment of the capsules was then completed.

Unlike a traditional Ferris wheel, the passenger capsules are not suspended but attached to the outside of the wheel. Aerodynamically shaped to minimize wind loadings, they are circular in cross section so they can rotate slowly within two mounting rings attached to the outer rim. They are driven by individual motors located in a void under the capsule floors and controlled by radio signals – which may be overridden by messages from an internal inclinometer. The tube structure supports an almost all-glass skin to maximize the view for passengers.

**Opposite:** The wheel is a circular truss with a triangular cross section with the capsules attached to the outer side. The horizontal tube in the foreground is not part of the structural system but one of 64 tuned mass dampers.

**Top:** The wheel being slowly hauled into position. Halfway through the lift the engineers had to stop for some weeks and carry out new calculations.

**Above:** Diagram of the wheel on the riverbank tied back to ground anchors.

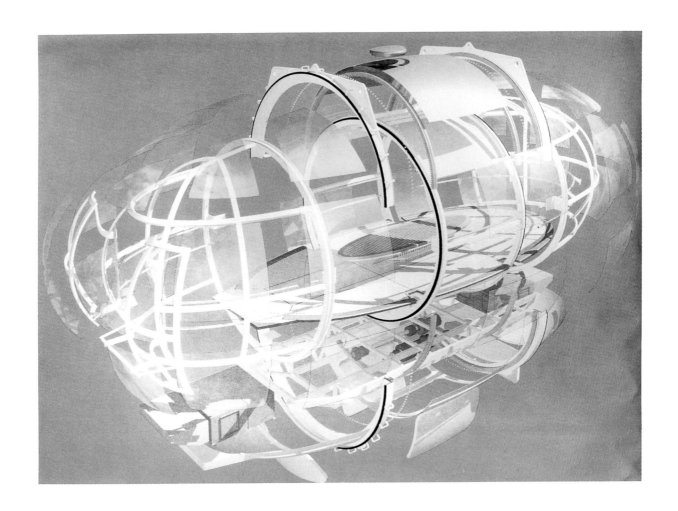

London Eye

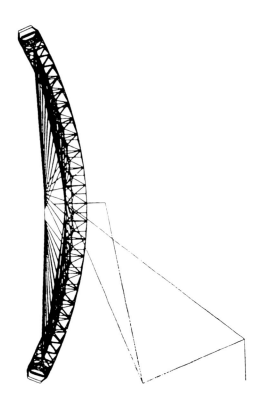

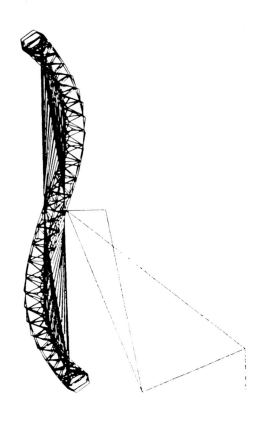

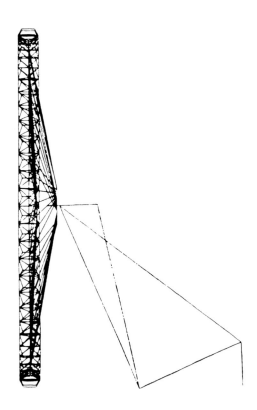

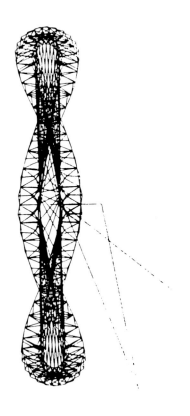

**Opposite:** Computer model and full sized prototype of a capsule showing the fixed ring attached to the wheel, within which the capsule is rotated by a computer controlled motor in the base.

**Above:** Engineers' diagrams of likely deformations of the wheel for which they had to design.

# Engineer: **Whitby & Bird with Specht Kalleja & Partners**

Stock Exchange and Chamber of Commerce, Berlin, 1998

## Architect: **Nicholas Grimshaw & Partners**

**Ludwig Erhard Haus accommodates the new Berlin stock exchange and chamber of commerce. It is the result of a competition won by British architects Nicholas Grimshaw & Partners with engineering consultants Whitby & Bird. The site is rectilinear on three sides and fragmented on the other. It is this irregular shape which in some ways set the visual agenda for the building, with its sweeping curve along one side and its three dimensional, armadillo-like form – a form which also owes something to Jean Balladur's daringly experimental concrete apartments of the 1970s at La Grande Motte on the south coast of France.**

The engineers, Whitby & Bird, and the architects looked at a variety of possible forms and structures, though the general shape of the original design and the clear ground floor remained more or less constant. One early idea was to deploy a series of lean-to hoops at the irregular side of the site and use butterfly trusses along the straight street façade demanded by the competition conditions. This turned out to be overly complicated and, after exploring a number of other ideas, the present ribcage solution was adopted. This is a series of 15 steel arches with varying spans and heights of up to nine storeys from which all the floors are suspended. Although it would have been physically possible to bridge some of the floor areas from side to side – and possible to incorporate column supports – the design team stayed with the suspension solution.

Among the engineering issues involved were progressive collapse (for which the engineers designed, though this is not required in Germany) and differential loadings on the arches in the areas of the two atriums.

The biggest and most interesting issue was movement. One of the consequences of hanging everything from the arches was that the whole building moved – the allowance at the lower floor levels was 50 millimetres (2 inches) in a day as the building was occupied and vacated. During its daily cycle the arches deflected, the hangers stretched and contracted and the floors deflected – all in a way quite different from the relatively restricted movement of a conventional structure. It meant that junctions between 'walls' and floors had to be designed to cope with movement and a special problem arose with the elevators. Eventually the team realized that they could use the technology which had evolved for elevators in ocean liners.

The engineers had originally selected high strength 75 millimetre (3 inch) macalloy suspension bars, but because of the long time required by the German building authorities before they would accept changes to the structural codes, these were changed to steel. Probably the biggest disappointment for the engineers was that instead of the visual excitement of slender rods supporting the floors in the two atriums, the suspension system is rather undramatic, with thick steel elements coated with various fire and weather layers.

**Preceding page:** One of the two great atriums from whose arches are hung the floors on either side.

**Below:** Despite its appearance as an apparently conventional office block, the street façade is actually grafted into the side of a radical structure.

Stock Exchange and Chamber of Commerce, Berlin

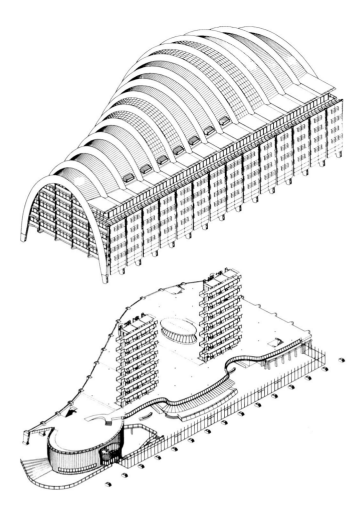

**Above:** The curved rear wall.

**Top:** The arches from which the floors hang are of different sizes but conform to the same geometry.

**Above:** Cutaway view showing how the elevator shafts are cantilevered up from the ground.

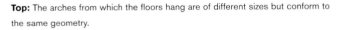

There were also some stability problems because the ground floor was clear and there was nothing vertical to act as bracing between ground and structure. This was also tied up with the issue of cladding the building. With such an irregular shape there was the good chance that every panel would have to be tailored individually and at great cost. The design team discussed many possibilities, including boat building technology, but eventually came to the idea of glass and a standing seam steel cladding – which could cope with the unusual geometry – laid over a concrete base. Apart from serving as a support structure for the external skin, the concrete acts as a shell, a structure which, like an egg, is stiff by virtue of its shape. Thus the engineering issue of stability was solved in the search for a sensible and economic skin structure.

Along the long straight Fasanenstrasse side runs a ground level promenade. This is the public zone from which people disperse, either to the office spaces above or down to the basement with its big conference facility. Two atriums slice into the building from the curving, undulating east to produce an E-shaped plan with natural perimeter light for all the floor areas, a plan refined and developed using solar modelling techniques. Although the building is meant to consume the minimum of energy, the final reduction is a relatively modest proportion of the norm for similar buildings. The amount of glass used was established as a balance between adequate natural illumination for all the office floors and energy-consuming artificial light. On the east façade, where the 'wall' curves down to around first floor level, the ratio of low emissivity glass to solid skin is 50:50. The other function of these sensational curving atriums is to allow the office spaces to be naturally ventilated.

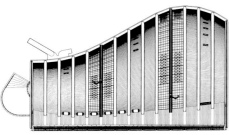

**Roof plan**

**9th floor plan**

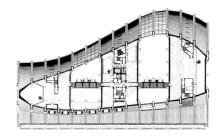

**8th floor plan**

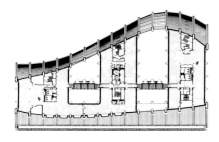

**7th floor plan**

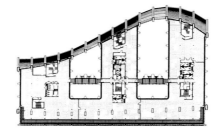

**6th floor plan**

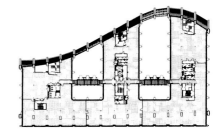

**5th floor plan**

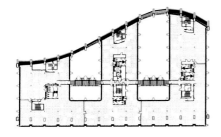

**4th floor plan**

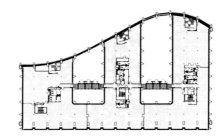

**3rd floor plan**

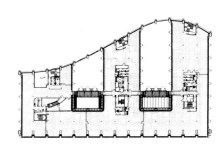

**2nd floor plan**

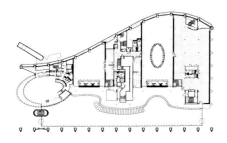

**1st floor plan**

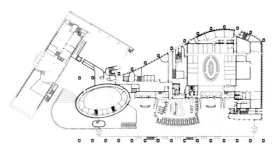

**Ground floor plan**

**Above:** Horizontal sections through the building showing the way the roof undulates, also the two atriums and their elevator bridges.

**Above:** The atrium with glass observation port for the exchange below. What look like columns supporting the edges of the floors are hangers suspended from the arches supporting the floors.

Whitby & Bird with Specht Kalleja & Partners

**Above left:** The elevator capsule rises on a vertical cantilever structure independent of the building.

**Above right:** Two continuous strips of glazing to one of the atriums (seen from the inside below left).

**Above left:** Atrium with heavy suspension rods supporting the edges of the floor slabs either side.

**Above right:** Conference room below the clear ground floor space.

Stock Exchange and Chamber of Commerce, Berlin

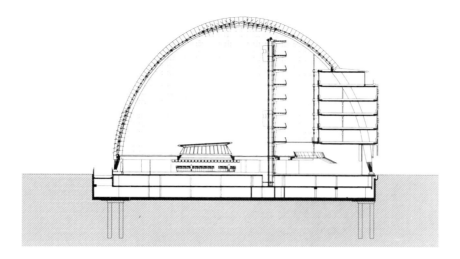

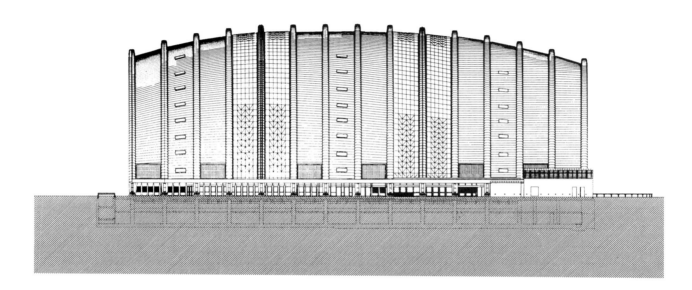

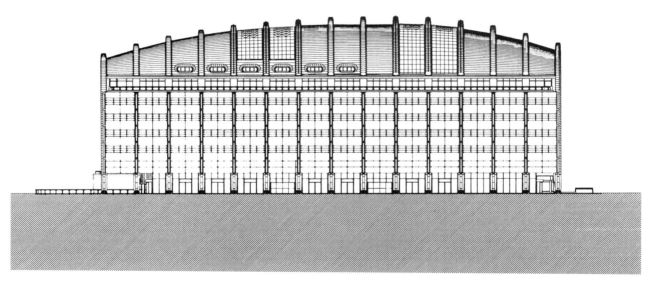

**Top:** Section across the building with, left, the stock exchange well in the open atrium, right of centre, a cantilevered elevator bridge tower and, right, the straight office block overlooking the street.

**Centre:** Rear elevation.

**Bottom:** Street elevation.

# Engineer: **Robert Nijsse of ABT**

Educatorium, Utrecht University, Utrecht, 1997

Architect: **Rem Koolhaas and Christophe Cornubert**

**This multi-functional building for the University of Utrecht has three exam rooms on the east, two lecture halls on the west and a 1000-seat restaurant on the ground floor, with student common rooms, some staff rooms and a big underground bicycle park. It is used for lectures and conferences as well as a meeting place for students – a learning factory, as OMA architect Rem Koolhaas describes it.**

The design's metaphor is of a huge concrete slab that has been folded up and over itself, maybe like a great book. The slope of the inner leaf accommodates the natural profile of the lecture theatre floors and enables a slanting ceiling for the restaurant beneath. The architects wanted a concrete structure cast on site in shuttering (timber moulds). Nijsse understood the architectural ethic that insisted materials be used as themselves, but he pointed out that shuttering would be very expensive and that the Dutch energy conservation laws ruled there be no cold bridges between the outer skin of buildings and the interior. This would have required ingenuity of a very convoluted kind, especially where the sloping concrete restaurant ceiling emerged into the outside to become a curving façade and doubled over itself to become the roof. Nijsse's contribution here was to construct the roll along the back of the lecture theatres as a half-cylinder steel frame with a layer of insulation covered

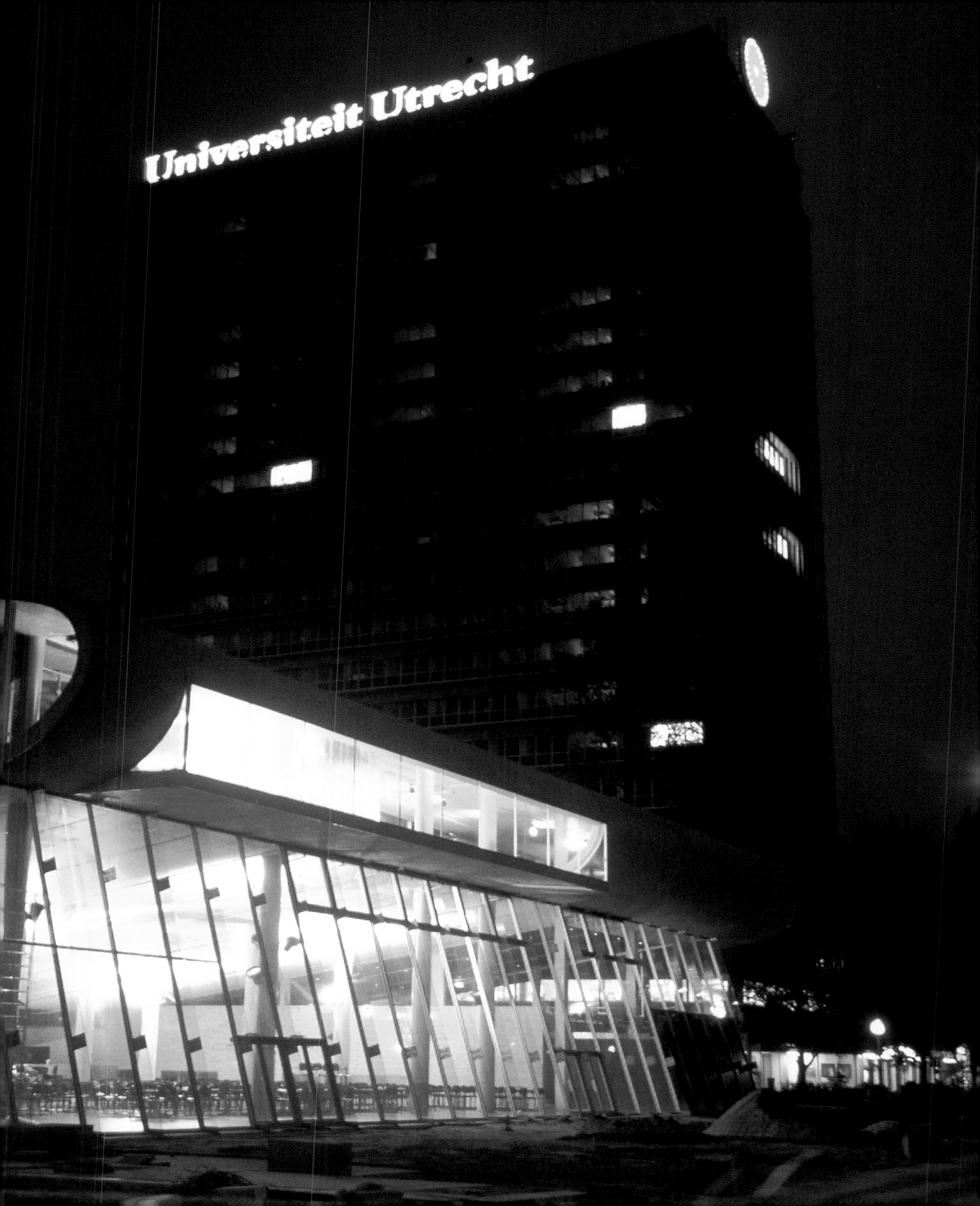
Universiteit Utrecht

**Preceding page:** The surprising structure of Utrecht's Educatorium which has further surprises half concealed and inside.

**Above:** The curving end of the building echoes the profile of the internal floors.

**Below left:** Egg-shaped intrusion at the back of the smaller lecture theatre.

**Below right:** Dramatic view showing the giant internal truss behind the students.

Educatorium, Utrecht University

by sprayed concrete. This was then trowelled to a smooth concrete-like finish. The inner curve was lined with raw plywood to form a seamless curving surface from floor round to ceiling, punctuated by a 25 metre (82 foot) wide window inserted in the curve by the architectural team late in the design.

The lecture theatres were big and the 500-seat auditorium had the special problem of a side glass wall, facing over parkland to the north, separated from the external curtain wall by a corridor/stairway. This rose from the entrance to the curved corridor at the rear of the lecture theatres. The architects wanted no columns in this elevation and Nijsse's solution was to devise a great storey-height truss set back the width of the corridor from the façade. He argued that this truss – which also acted as the support structure for the glass wall to the lecture theatre – should express very plainly the considerable stresses involved in supporting the load over a span of 35 metres (115 feet). This stark engineering solution allows the forces to be read from the construction.

The roofs of the lecture theatres had to span 20 metres (66 feet). On the smaller of them the engineers experimented with doubled folded steel plates, followed by triple plates and added concrete to form a composite structure with pre-stressed reinforcement with arabesque undulations. In the end the conventional solution was adopted of a series of beams, steel plate and concrete. However for the auditorium a radical and unprecedented solution was chosen. The architects insisted that the edge of the roof should be no more than 400 millimetres (16 inches) deep in order to match the floor slab edge – preserving the notion of the building as a single slab folded over itself. Nijsse reports dryly, 'This was virtually impossible and made us consult with the architects on a frequent basis.' Nijsse needed a depth somewhere between 500 and 600 millimetres (20 and 24 inches). During a collaborative brainstorming session a structural innocent asked why so much depth of concrete was needed. The conventional answer is that in reinforced concrete the steel reinforcing, acting entirely in tension (the concrete acting almost entirely in compression), has to be covered by the concrete. But the group suddenly realized the steel does not have to be covered. This would mean eliminating a great deal of very heavy concrete and reduce the dead load the roof structure would have to carry. And so Nijsse's team, using a powerful finite element computer program, evolved an incredibly elegant solution. They designed a thin concrete slab which takes all the compressive forces in the roof, while emerging from the bottom of this slab is a network of reinforcement bars that form the skeleton profile of a very shallow hemisphere in the lecture theatre ceiling. There were particular problems in setting up the formwork, which Nijsse solved using a mixture of standard and prefabricated shuttering and by welding secondary bars while the shuttering was being taken away.

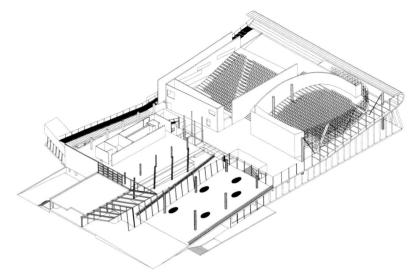

**Above:** Middle layer of the Educatorium with two lecture theatres, right, over the student cafeteria.

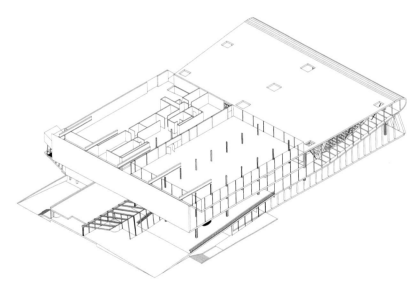

**Above:** Upper level with examination rooms, left, and the roof of the lecture theatres right.

**Above:** Lecture theatre with exposed catenary tension reinforcement allowing the concrete slab from which it emerges to be as thin as possible.

Educatorium, Utrecht University

**Above left:** The ramp on this glazed elevation continues the slope of the lecture theatres.

**Above right:** Connection of the new building and the old.

**Above:** The all-plywood interior of the curved wall has alternative uses.

# Engineer: **Conzett Bronzini Gartmann**

## Traversina and Suransuns footbridges, Viamala Gorge, 1996 (destroyed 1999) and 1999

The Traversina Bridge, spanning 48 metres (157 feet) across a remote ravine in the Viamala Gorge in Switzerland, is a tour de force in timber and cable structure. Its first design criterion was that the major elements should weigh no more than 4.3 tonnes (4.7 tons), the maximum load of the most powerful available transport helicopter. The two elements were the parabolic truss and the deck with its stabilizing H frames. The truss is made up of 23 quite complicated triangular frames located at intervals at right angles to the span and increasing and diminishing in size from one side to another.

Their tops are connected by a wide strut of glue laminated (glulam) larch to carry the deck and provide lateral stiffness against wind loads. The two bottom corners of all the triangles are linked by cables on either side. The parabolic profile of these two cables is that of the structure's bending moment diagram for a uniformly loaded condition. The struts are further located in space by criss-crossing diagonal cables. When the bridge is loaded by the deck, the H frames and people, the cables are all in tension and the timber struts are mostly in compression.

**Above, left and right:** Joining timber and cable, two very different materials, involved complex geometry and called for some engineering ingenuity.

**Opposite, above and below:** Although the Traversina Bridge is a timber structure, it is actually a very pure design in which timber is used for compression struts and cables for tension members.

**Preceding page:** The intricate timber and cable structure of the Traversina Bridge was brought in as three elements by helicopter.

**Below:** Side elevation of the Traversina Bridge.

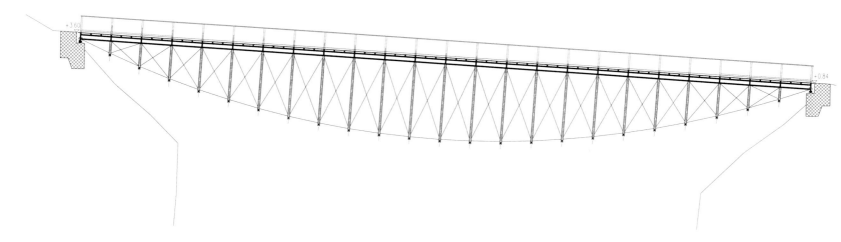

Traversina and Suransuns footbridges

**Above:** View of the underside of the walkway which sits on the long piece of timber at the apex of the A legs. The cables are the tops of the diagonal bracing between the legs.

**Right:** Engineer's drawing of a typical leg indicating junction plates between paired legs.

**Opposite:** The Traversina Bridge in use.

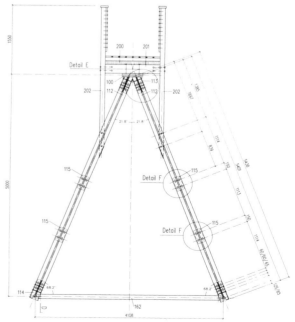

Traversina and Suransuns footbridges

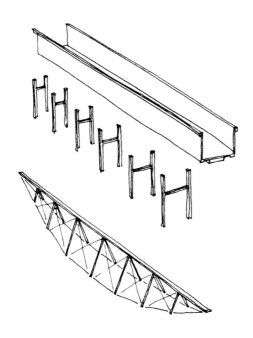

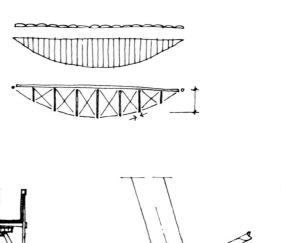

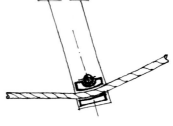

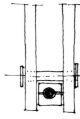

**Above left:** The three elements of the Traversina Bridge: walkway, H format stabilizers and the main timber truss.

**Above right:** The truss and walkway were designed with the helicopter-based installation in mind: they had to have their own structural integrity.

The sides of the walkway are in solid timber, partly to reassure users and partly for structural reasons. With the deck they also form a channel profile which helps to keep the walkway level under differential loads. This shape was also necessary if the deck structure was to remain together during the helicopter lift. The channel is cradled by H frames whose legs lock into the structure of the triangular frames beneath. The end bearings are steel rollers on timber blocks.

The timber structure was pressure treated unless the relevant timbers – such as the deck and balustrades – can be replaced in the future without imperilling the integrity of the structure.

**The Suransuns footbridge** in the same region is a 40 metre (131 foot) ribbon bridge over the Hinterrhein River with a footway less than 1 metre (3 feet) wide. It drops 4 metres (13 feet) from one side of the riverbank to the other and has a sag of only 1 metre (3 feet). In contrast with the visual complexity of the Traversina, the Suransuns is breathtaking in its clarity and simplicity: a deck of granite slabs in which are planted the uprights of the delicate steel handrail which simply spans across the river in a very shallow catenary. It seems impossible that this almost straight rock and steel assemblage could ever stay up – least of all accommodate moving pedestrians. The only trace that something else is going on are the pieces of steel emerging from under either side of the last slabs on each approach. These are attached to the concrete abutment in an elegantly simple and reassuringly visible arrangement of steel parts.

Closer examination reveals little extra. Underneath, two pairs of 15 x 60 millimetre (3/4 x 2 1/3 inch) high strength steel straps carry the 1100 x 250 x 60 millimetre (43 x 10 x 2 1/3 inch) granite slabs of the deck. These are the suspension cables which are anchored to the heavy abutments which themselves are tied back to the bedrock by rock bolts. The end of each slab is drilled for the threaded ends of two 16 millimetre (3/4 inch) diameter handrail uprights. There is a locating nut on top of the slab and underneath the thread passes between the paired straps and a metal cross plate against which the bottom bolt is tightened. The handrail on top is a 10 x 40 millimetre (3/4 x 1 3/4 inch) steel tube. Between each of the slabs is a 3 millimetre (1/10 inch) joint of medium-hard pure aluminium which accommodates movement and protects the edges of the stone slabs as they are compressed together by pedestrians walking over the bridge.

In some ways this structure acts like a beam in which the upper zone is in compression and the lower in tension: steel performing best in tension and stone best in compression. The other analysis is that the stone slabs act as if they were the voussoir stones of an upside down arch, the load created by the suspension straps keeping them together. This is engineering of irreducible clarity and simplicity in which the few elements and details are no more and no less than they could possibly be – which is how the great Renaissance architect Alberti defined the inner core of beauty.

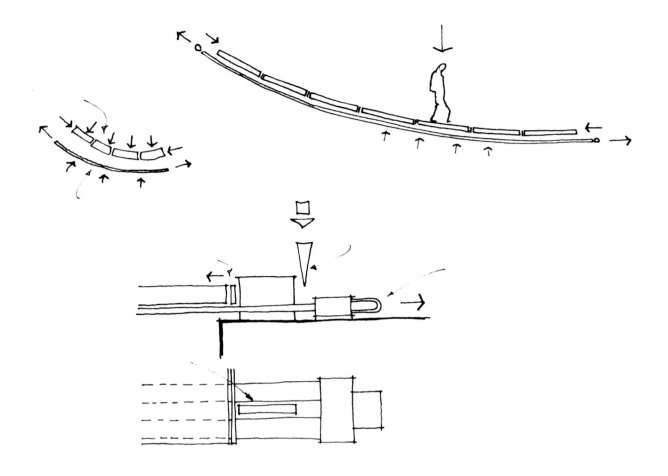

**Top:** Structural diagram explaining the way the forces work in the Suransuns Bridge.  **Above:** The anchoring system in elevation and plan.

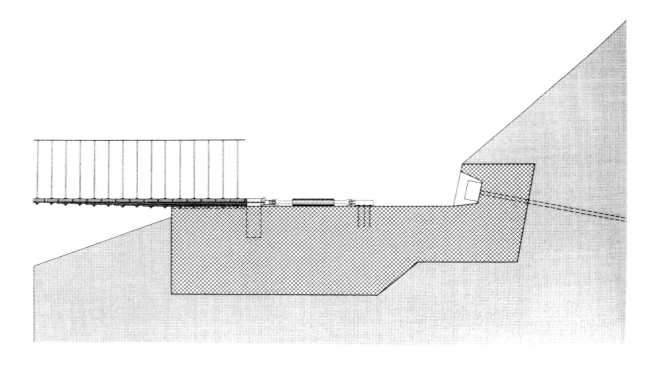

**Above:** The very high tension forces in the Suransuns Bridge mean that it has to be anchored securely – here by a heavy concrete block tied back to the bedrock with big rock anchors.

**Below left:** The bridge appears simply to come to a halt.

**Below right:** A stone walkway, delicate steel handrails: extraordinary simplicity.

Traversina and Suransuns footbridges

**Above:** The underside of the Suransuns Bridge: the handrail rods are held on top of each slab by a locknut. A bottom locknut secures the rod to a metal crosspiece, thus clamping the two straps.

**Below:** A sectional elevation showing how the structure actually extends into the hills on either side.

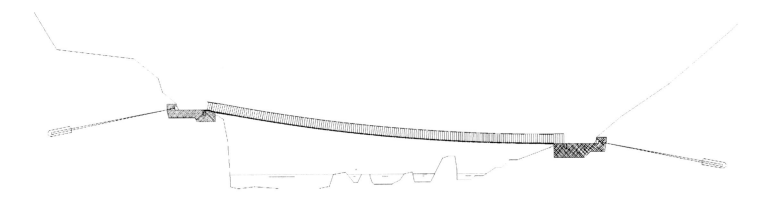

# Engineer: **Marc Mimram**

## Solférino Bridge, Paris, 1999

**This new two-level bridge crosses the Seine in a single 106 metre (348 foot) span from near the Musée d'Orsay on the Left Bank to the Tuileries Gardens and the nearby Louvre on the Right Bank. From a distance its form appears relatively conventional: a long, low, trussed arch with a flat footpath on top linking roadways on either side. In fact the lower chord of the arch truss carries an additional pedestrian route in the form of steps following the curve of the arch which emerge through a long slot in the middle of the wide flat deck.**

This lower route, narrow and curving, starts on the south side at a busy low-level auto throughway and on the north at a riverside promenade with subway access to the Tuileries Gardens. Pedestrians can start at riverside level on one side and reach the embankment or road on the other.

The bridge was the outcome of a competition. Marc Mimram's winning design was originally a very light aluminium structure. But aluminium is more expensive than steel and less flexible, so eventually the bridge was built entirely in steel. Mimram, an architect and engineer, finds inspiration in natural forms. This bridge is derived from the structure of the human spine, with a pair of backbones made of cast steel vertebrae between which pedestrians climb up from the lower levels.

The basic structure is two steel arches joined by transverse ribs. The lower chord of the arch has the formation of a curving horizontal ladder on which sit the Vs of the supports for either side of the upper deck. In diagrammatic terms the upper deck is treated virtually as two independent decks supported off the main ladder arches on either side.

At the springing of these arches at either bank the V props are set at a steep angle, but as the arch rises they diminish in height and their angle flattens out more and more in a kind of static Olympic wave. The ladders turn out to be more complicated because they are actually the outer one of paired lower chords which rise from the river bank abutment as a single unit and separate to support the lateral supports for the V props. In addition, the two sets of ladder arches are tied together below the steps and, overhead, the cantilevering extensions of the ties form joists for the finely grooved azobe hardwood decking.

**Preceding page:** The bifurcating arches of the support structure actually carry a secondary stepped walkway.

**Above:** Computer model of the bridge showing the paired V struts either side of the secondary walkway supporting the upper platform.

**Below:** Plan of the bridge across the Seine. The Tuileries Gardens are to the left and the Musée d'Orsay to the top on the right.

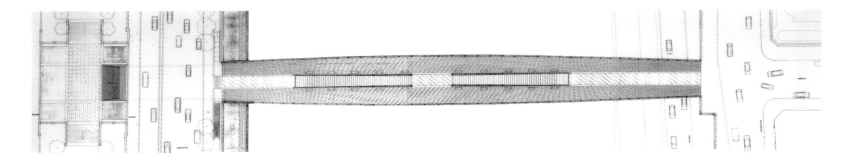

Here is a structure of skeletal elements, whose inherent complexity is enhanced by the unexpected introduction of the lower, stepped pedestrian route, with its slanting steel handrails and steps adding to the visual richness. It needs to be said that though it is possible to discuss the elements of the bridge in terms of struts and ladder arches and supporting props, this is not just a complicated shallow arch supporting an overhead footpath with additional lateral restraints. The components of the bridge are all essential elements in the three dimensional engineering ensemble.

Like London's quite different Millennium Bridge, which oscillated sideways under pedestrian traffic and had to be closed as soon as it opened, the Solférino Bridge had its swaying problems. These were quickly and quite inexpensively resolved by installing 25 tonnes (27.5 tons) of tuned mass dampers, which cancel out the swaying caused by pedestrians falling into step in harmony with the bridge's natural frequency. For some time the website of Ove Arup, the British engineers responsible for the Millennium Bridge, hinted somewhat unconvincingly that their difficulties partly stemmed from the fact that Mimram had not published his analysis of the problem and its solution.

In a neat twist of fate, the constructors of this quasi homage to the great cast iron engineering-architecture of nineteenth century Paris were the Eiffel Company, the lineal descendants of the company which built Gustave Eiffel's great tower a short distance downstream.

**Below:** Engineer's drawings detailing the lower stepped walkway which emerges towards the middle of the upper platform.

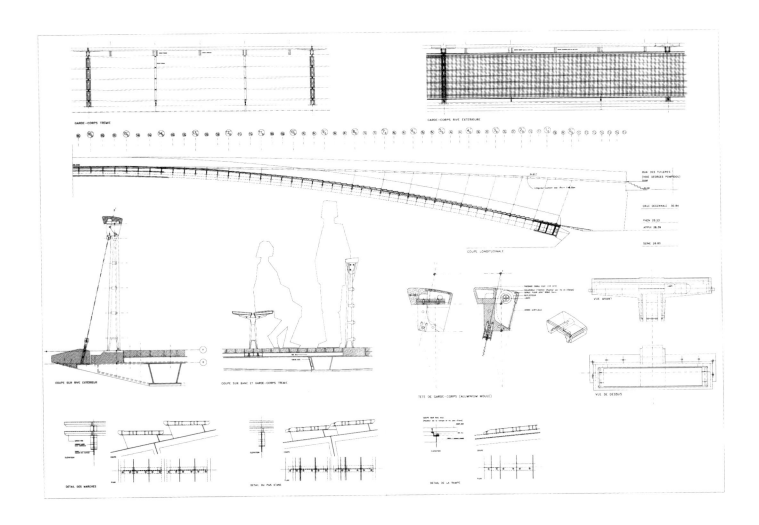

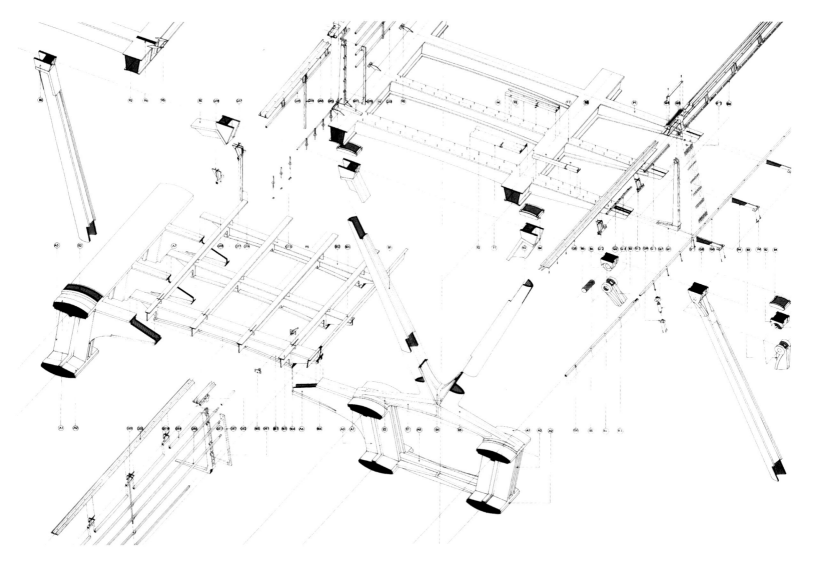

**Above:** Exploded diagram of the bridge's sculptural prefabricated components.

**Below:** The arch across the Seine has classic proportions and form. What is so different is the use of the arch as a secondary walkway system.

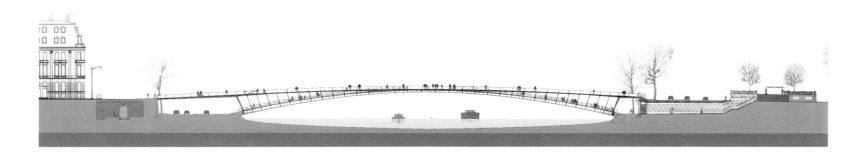

Solférino Bridge

**Above:** The bridge from the Left Bank.

**Below:** Diagrammatic analysis of the bridge's initial tendency to wobble in harmony with its natural frequency with the use of spring dampers to counteract the oscillation.

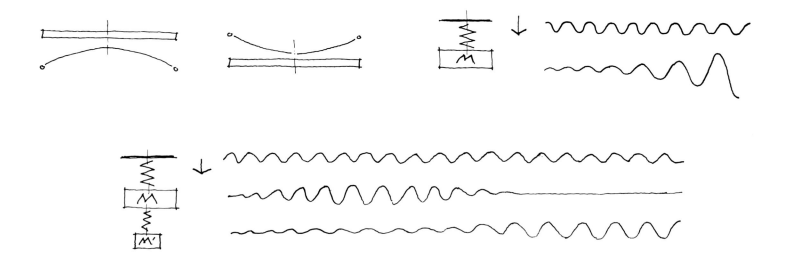

**Above:** Computer image of the bridge.

**Below:** Three dimensional structural diagram.

**Opposite:** Sections through the bridge at various stages from mid span to ground level showing how the V struts change in shape and angle.

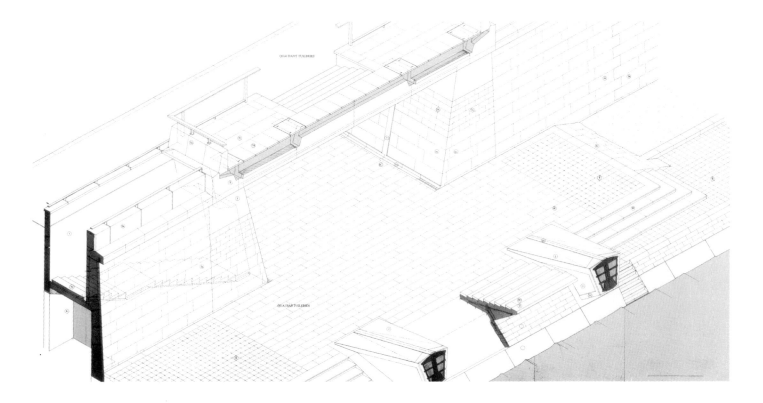

Solférino Bridge

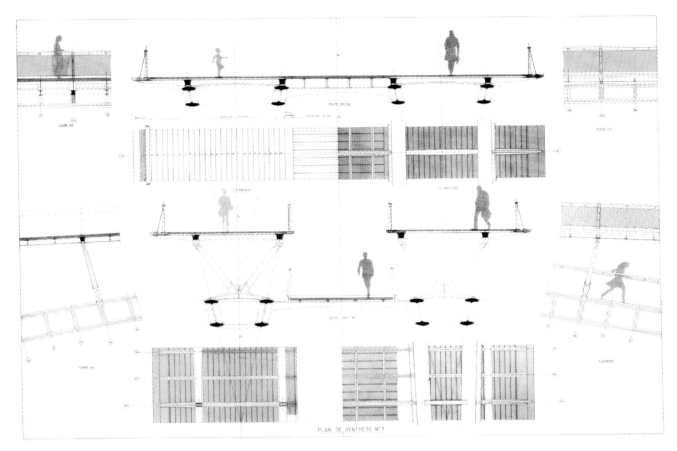

PLAN DE SYNTHESE N°3

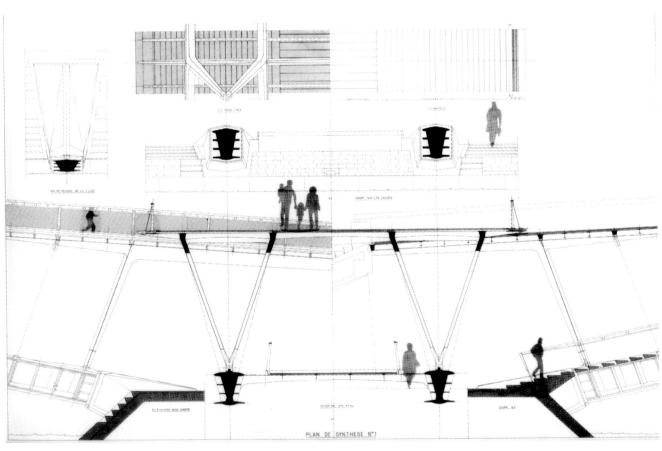

PLAN DE SYNTHESE N°1

# Engineer: **Julius Natterer**

Expo Roof, Hanover, 2000

Architect: **Herzog + Partner**

**The Expo Roof was designed to remain as a permanent marker at the centre of the Hanover Expo 2000. During the show it served as a shelter for the public, for performers and for small pavilions and restaurants. The ten roofs in the shape of umbrellas are hung at a height of 26 metres (85 feet) on timber towers. Each shell covers an area 40 metres (430.5 feet) square and the ten shells are all connected so they are statically dependent on each other.**

What is particularly interesting is that they are made from glue-laminated timber (glulam) – a special area of expertise for engineer Julius Natterer, and for the architects as well. There is a tendency to think of timber as an inherently lightweight form of structure. This is a heavy timber structure. Quite apart from its bravura and structural daring, its area of 16,000 square metres (172,000 square feet) is of unprecedented scale and size.

   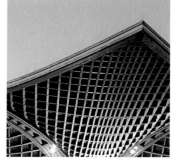

Each umbrella is made up of five elements: the four 10 to 15 metre (33 to 49 foot) deep piles of the foundations supporting, via steel fittings, the triangulated timber tower structure. A steel pyramid sits on top of the tower to act as anchorage for the four timber cantilever beams which support two sides of each of the double curvature grids. The roof is covered with a waterproof membrane held 50 millimetres (2 inches) off the timber structure by cables following the longitudinal ribs. This allows any wet timber to dry out and lets air permanently circulate around all the timbers.

The four 16 metre (52 foot) long legs of the towers are debarked silver fir trees with diameters of around 700 millimetres (28 inches), mounted upside down so the broader girth is at the top where the stresses are higher. Some of the trees were as much as 200 years old. The bark was peeled off using high pressure water and then, because the trees were so big, they were sawn down the middle to help them dry out over a seven month period before construction. Additional stiffness is provided by triangular webs made of diagonal glulam planks faced with a 33 millimetre (1 1/3 inch) thick lumber veneer. These slot between the two halves of the former tree and are held in place by dowels and bolts with spring connectors to cope with shrinkage as the timber dries out. As it was, the seven months drying time turned out to be insufficient to bring the timber to an acceptable moisture content, so the allowable axial stress had to be reduced by one third and the stiffness factor by one sixth.

An open 32 tonne (35 ton) steel cap sits on top of the tower to which the four timber cantilever trusses are attached. They are 3 metres wide and 19 metres long (10 by 62 feet) with a depth at the attachment point of 7 metres (23 feet). Although only the lower chords are visible their structure is very complex. It involves the use of what are called K-diagonals which segue into a box girder at the outboard end where they attach to the edge girder of the shells. The shells are made up of criss-crossing ribs on a grid which varies from 380 millimetres (15 inches) in areas of high stress to 1600 millimetres (63 inches) at high points on the shell. Ribs are made up of between eight and ten 30 millimetre (1 inch) thick layers, which are screwed together and interleave at cross joints where they are also bolted. Rib layers are glued together only at points of high stress. These are the main elements visible from the ground but there is another layer of structure, this time a grid of a much closer, 100 millimetre (4 inch) frequency laid at 45 degrees to the direction of the lower grid. The shells were assembled over three months in a nearby exhibition hall, but the cantilever trusses were fabricated 700 kilometres (435 miles) away and had to be escorted along closed roads at night to the site.

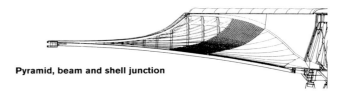

**Pyramid, beam and shell junction**

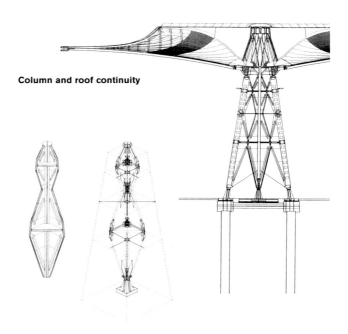

**Column and roof continuity**

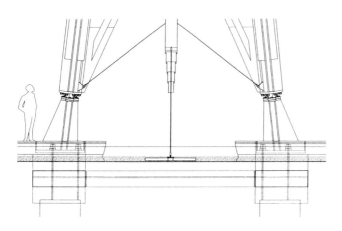

**Preceding page:** Like giant basketweave Natterer's vast roof is made up of interlinked warped surfaces.

**Above:** Foot detail.

**Right:** Column and roof elements.

**Column dissected: timber left, steel right**

Although Julius Natterer has a number of daring timber structures to his credit, such a cutting edge design as this called for wind tunnel testing and analysis in terms of wind induced oscillations. The results were unexpected: wind loads, for example, turned out to be down rather than up. Because of the sheer size and complexity of the structure, involving several thousand nodal points and nine thousand structural members, the only possible mode of analysis was via computer and this involved more than 150 different analysis and calculation stages.

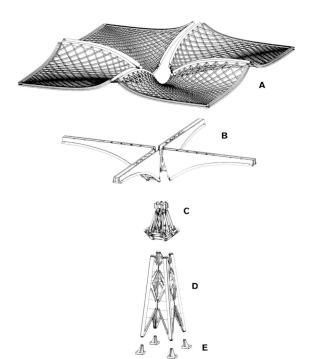

**Above:** The complete ensemble of umbrellas.

**Left:** The elements of each of the umbrellas: pile caps supporting the braced tall tree trunk pyramids, steel support cap for the timber support cantilevers and the double layer timber of the roof.

**A:** Skin
**B:** Cantilevered beams
**C:** Steel pyramid
**D:** Tower
**E:** Feet

# Engineer: **Paul Muller with RFR**

Hall F, Charles de Gaulle International Airport, Terminal 2, Paris, 1997

Architect: **Paul Andreu, ADP**

**This is the third arrival-departure module in the chain forming Terminal 2 of Charles de Gaulle International, the main airport of Paris. Architect Paul Andreu conceived it as a much larger and more fluid version of the two satellites making up the second terminal he had designed 20 years before. Here the aircraft nosed into air bridges at the perimeter of part-circular structures.**

These were arranged on either side of a spine which acted as a route for the movement of traffic and people. The new module was conceived with a hollow oval plan with two 200 metre (656 foot) long 'fingers' on each side. Up to 14 aircraft nose in to air bridges attached to the sides of each finger. At the moment half of the new oval has yet to be built.

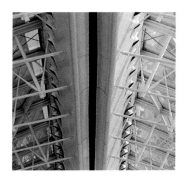   

Andreu began his concept designs in the early 1990s. He worked with a group of architects and engineers that included the late Peter Rice and his practice RFR, who were to be responsible for the most visible parts of the design, the structural steelwork and the façades. Andreu conceived the main body as a concrete structure which was to enclose the essential infrastructure: primarily passenger circulation, but also rail and roads.

This structure is clad in zinc and at intervals Andreu has cut out large oval shapes whose glazing reveals a heavy grid of small windows punched through the curving concrete shell which underlies the main roof. The two fingers emerge dramatically from this organic shape as all-glass roofs whose cross section is close to that of a gothic arch. Above the glass is a brise-soleil of perforated stainless steel – which turned out to have a better performance than the fashionable option of fritting glass with small opaque dots. The sunshades form what the design team has described as elytrons, from the name for the paired wing cases of a flying beetle. On the outside of the glass skin the struts that support the elytrons are located directly in line with the compression struts of the interior trusses. Thus the glass appears to occupy a layer between the apparently luminous metal of these sunshades and the fuzzy boundary formed by the lower chords of the trusses and the ends of the compression struts – which were designed to be lit at night by fibre optic dots.

**Preceding page:** Dramatic computer image of the curving, spoked side wall of a finger. Beyond, the cut-out in the wrapping for the concrete main structure.

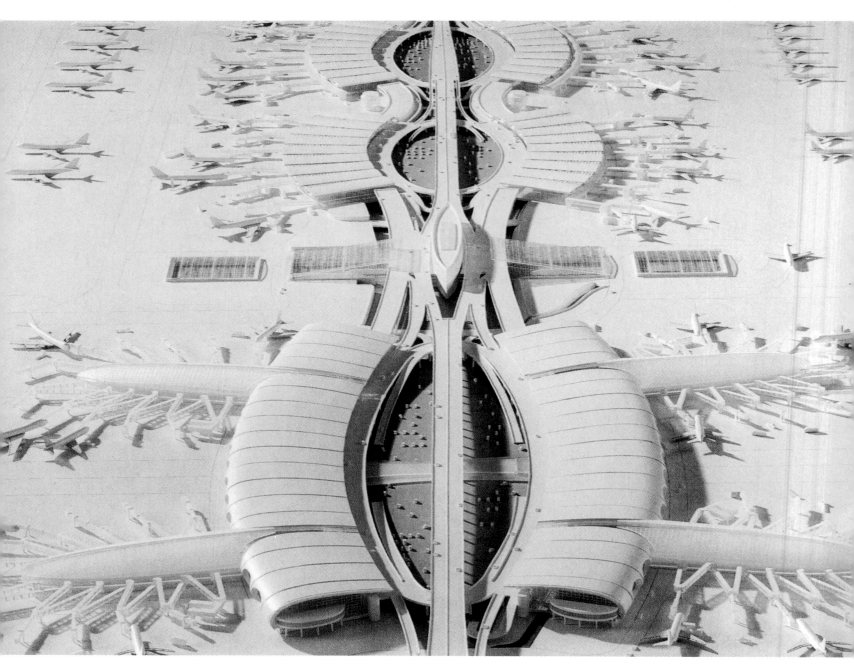

Hall F, Charles de Gaulle International Airport, Terminal 2

**Opposite:** Model of the airport development with two existing modules already built. Half the big oval of the new module in the foreground has been built.

**Above:** Concept drawing of the form of a finger.

**Below:** The overhead geometry of the junction between a transparent finger and the main arrivals and departure structure.

**Above, left and right:** Views of the roof structure of a finger. The 'knife' appears to provide a long cantilever's support for either side of the roof. In fact the trusses are arches and the knife disguises some structural untidiness and environmental controls. The glazed roof is protected externally by lightweight sunshades fixed to the same struts which form the vertical elements of the roof trusses.

**Above, left and right:** Views of the elaborate geometry of the curved side walls to each finger.

Hall F, Charles de Gaulle International Airport, Terminal 2

The apexes of the delicate arches appear to be connected by what the designers describe as the 'knife'. This serves as a kind of visual backbone which emerges from the main concrete structure and continues almost to the end of the finger in an apparently impossible downward-curving cantilever. But it is not quite what it seems. The arches of the walls and roof are not hinged centrally at the knife: they are actually two sides of the same rigid truss which at the widest point of the finger involves a span of up to 50 metres (164 feet). Towards the end of the finger, however, the two sides are not necessarily in the same vertical plane and the resultant horizontal forces are accommodated by truss rods inside the knife. The structure of the knife also accommodates other geometric difficulties and unsightliness. Andreu says he was 'glad we did not let ourselves become trapped by a detail that would have belied the spirit of the building... As on several other occasions, we deliberately turned our backs on whatever we regarded as purely a matter of technological rhetoric.' The other functions of the knife are to provide smoke outlets, environmental services and interior lighting for the finger.

The lower edges of the glass roofs roll down and under the fingers in a semicircle, echoing the curves of the main structure. From outside they appear to be seated on a long concrete plinth. In fact the lower ends of the trusses turn into solid semicircular beams, which are partly supported on the plinth and partly by a series of fan-shaped structural arrays. The fan is formed from tension rods attached to the ends of horizontal masts fixed at right angles along the edge of the upper Departures concourse – radiating out like spokes to the circumference of the half circles of steel. They give the upper concourse the quality of what the designers call a 'flying carpet', with the glass roof magically disappearing out and down far beyond the edges of the floor. Arriving passengers have an exciting experience with the arrays of horizontal struts and radiating spokes above and to the side revealing the busy activities of the airport beyond. Andreu says, '...obviously it is not flying... But the form of the structures that connect it to the metal frame makes it look as if it were isolated in space, as if it were sustained rather than sustaining. At the origin of this form was our desire to put some distance between the passengers and the façade. Placing the façade out of reach made it easier to keep it clean. Moreover, this tactile distance made the glass envelope more remote and mysterious.'

**Below:** Diagram showing how the structure is supported from horizontal masts attached to the edge of the upper concourse for departing passengers.

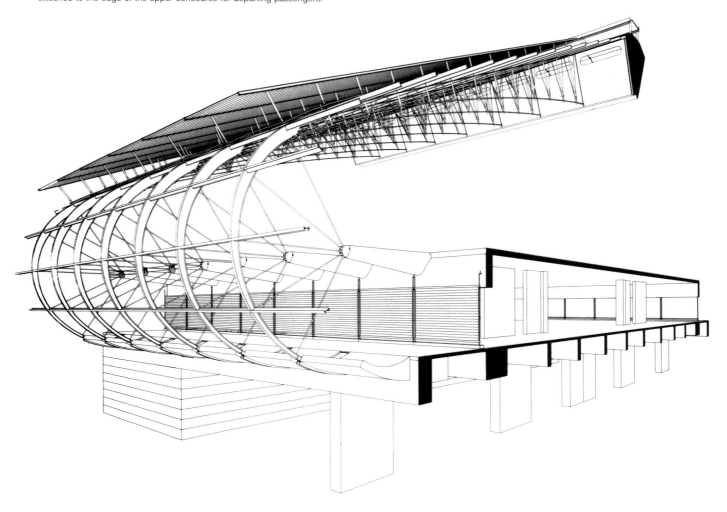

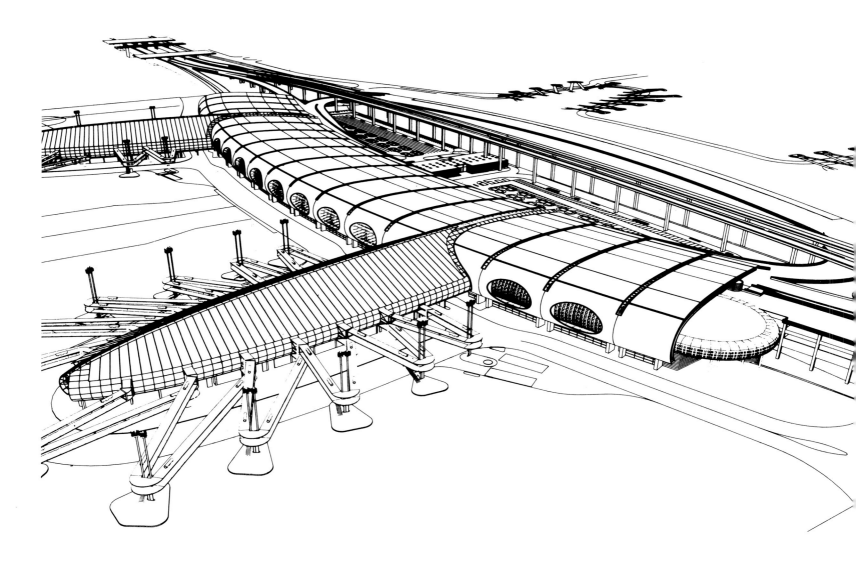

**Above:** Drawing showing the relationship of the fingers to the main structure.

**Below:** The finger is actually part of a symmetrical array: the main building's profile
is outlined on the left.

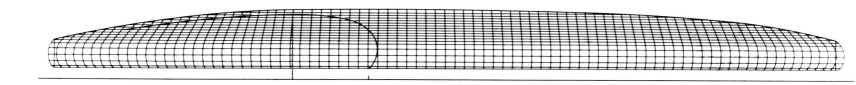

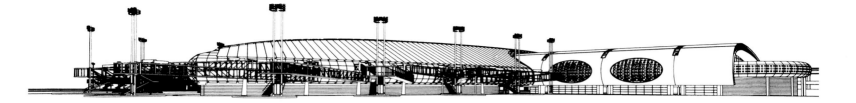

**Above:** Perspective showing a finger and the wrapping of the main building with a skin which has large apertures to allow the inner perforated concrete structure to emerge.

**Below:** Inside one of the fingers looking back with the structurally enigmatic knife overhead.

# Project Credits

## Park Keeper's Flat and Public Lavatories, Shungu-cho, Hyogo Prefecture

**Structural Engineer:** TIS & Partners

**Architect:** Shuhei Endo Architect Institute

**Client:** Hyogo Prefecture

**Construction:** Inoue Building Firm

**Corrugated Construction:** Nakamitsu Kenzai Co., Ltd

## Millennium Dome, Greenwich, London

**Structural & Services Engineers:** Buro Happold Consulting Engineers

**Architect:** Richard Rogers Partnership and Imagination Ltd

**Client:** New Millennium Experience Co, Millennium Commission.

**Construction Manager:** McAlpine/Laing joint venture

**Planning Supervisors:** Ove Arup & Partners (on behalf of RRP)

**Fire Engineering Consultants:** Buro Happold/Fedra

**Environmental Consultants:** Battle McCarthy

**Landscape Architects:** Desvigne & Dalnoky/Bernard Ede

**Fabric Roof:** Birdair Inc

**Main Steel Structure:** Watson Steel Limited

**Fire Consultants:** FEDRA

**Civil & Remediates:** WS Atkins

**Specification Writers:** Schumann Smith

**Acoustic Consultant:** Sand Brown Associates

## Japan Pavilion, Expo 2000, Hanover

**Structural Engineer:** Buro Happold

**Architect:** Shigeru Ban

**Client:** Japan External Trade Organization (JETRO)

**Structure Consultant:** Frei Otto

**Contractor:** Takenaka Europe BmbH

**Construction:** TSP Taiyo PERI

## Eden Project, Bodelva, Cornwall

**Civil and Structural Engineers:** Anthony Hunt Associates

**Architect:** Nicholas Grimshaw & Partners

**Geotechnical Engineers:** John Grimes Partnership

**Landscape Architects:** Land Use Consultants

**Mechanical & Electrical Engineers:** Ove Arup & Partners

**Cost Consultant:** Davis Langdon & Everest

**Project Manager:** David Langdon Management

**Planning Supervisor:** Aspen Burrow Crocker

**Constructor:** McAlpine Joint Venture

**Biome Frame:** Mero UK

**Bione Cladding:** Foiltech

**Link Frame:** Snashalls Steel Fabrications

**Visitor Centre Frame:** Pring & St Hill

**Foundations:** Dean & Dyball

## Great Glasshouse, National Botanic Garden of Wales, Carmarthenshire

**Structural Engineer:** Anthony Hunt Associates

**Architect:** Foster & Partners

**Client:** National Botanic Garden of Wales

**Landscape Architect:** Gustafson Porter

**Landscape Architect:** Colvin & Moggridge ,

**Quantity Surveyor:** Symonds Group

**Mechanical & Electrical Engineer:** Max Fordham & Partners

**Construction and Project Manager:** Schal

**Concrete Structure:** Byrne Brothers

**Steelwork:** Watson Steel

**Roof Glazing:** Metallbau Fruh

**General Builders:** Cowlin Construction

**Precast Conctrete:** Histon Concrete Products

**Laminated Glass:** Bruder Eckelt

## William Hutton Younger Dynamic Earth Centre, Edinburgh

**Engineer:** Ove Arup & Partners
**Architect:** Michael Hopkins & Partners
**Client:** Lothian and Edinburgh Enterprise Ltd. (LEEL)
**Quantity Surveyor:** Gardiner and Theobald
**Project Management:** Kier Project Managament
**Management Contractor:** Laing Management (Scotland) Ltd
**Leisure Consultant:** Grant Leisure Group, Acoustic Consultant: Robin
**Mackenzie Partnership, Exhibition:** Event Communications Ltd.,
**Masonry:** Dunhouse Quarry Co Ltd, Fabric Roof Consultants: Birdair Inc,
**SH Structures Ltd., Land Engineering, Fit Out Contractor:** Swift Horsman, Shop Fitout; Skakel &  Skakel, Metalworks: James Blake and Co. (Engineers) Ltd
**Lighting;** Guzzini Illuminazione UK Ltd, Bega from Concorde Lighting Ltd, Bar/Café Furniture: Amat through Howe Projects Ltd
**External Paving:** Northowram Hills & Stanton Moore Stone by Land Engineering
**Landscape:** The Paul Hogarth Comany

## Rose Centre for Earth and Space, American Museum of Natural History, New York

**Structural Engineer:** Weidlinger Associates, Inc.
**Architects:** Polshek Partnership Architects
**MEP, Fire Protection:** Altieri Sebor Wieber
**Civil Engineering:** Joseph S. Loring & Associates
**Landscape Consultants:** Kathryn Gustafson, Anderson & Ray, Judith Heintz
**Specifications:** Construction Specifications Inc.
**Acoustics:** Shen Milsom & Wilke
**Code:** Jerome S. Gillman Consulting Architect P.C
**Curtainwall:** Heitmann & Associates Inc.
**Tension Structure Consultant:** TriPyramid Structures Inc.
**Vertical Transportation:** Van Deusen & Associates
**Color Consultant:** Donald Kaufman Color
**Graphics Consultant:** Pentagram Design
**Building Preservation:** Archa Technology
**Water Features:** Gerald Palevsky
**Lighting:** Fisher Marantz Stone

## Tokyo International Forum

**Engineer:** Structural Design Group
**Architect:** Rafael Viñoly Architects
**Client:** Tokyo Metropolitan Government
**Mechanical Engineer:** P.T Morimura & Associates
**Acoustic Engineer:** Yamaha Acoustic Research Laboratory
**Theatre Mechanic and Stage Lighting:** Theatre Workshop
**Lighting Design:** Claude R. Engle, Lighting Consultant
**Quantity Surveyors:** Futaba Quantity Survey

**General Contractors:** Taisei Corporation (theatre buildings), Obayashi Corporation (glass hall buildings)
**Steel Manufacturers:** Kawasaki Heavy Industries, LTD, Kawagishi Industries, LTD.
**Glass Manufacturers:** Asahi Glass Company
**Stage Machinery:** Mitsubishi Heavy Industries Ltd.

## Mercedes Benz Design Centre, Sindelfingen

**Engineer:** Ove Arup & Partners, IFB Dr Braschel & Partner GmbH
**Architect:** Renzo Piano Building Workshop Architects
**Client:** Mercedes-Benz AG
**Services Engineers:** FWT Project & Bauleitung Mercedes Benz AG
**Acoustics:** Müller BBM
**Interiors:** Santolini

## Lords Media Centre, London

**Engineers:** Ove Arup & Partners
**Architects:** Future Systems
**Client:** Marylebone Cricket Club
**Services Engineers:** Buro Happold
**Cost Consultant:** Davis Langdon & Everest
**Fire Engineering:** Arups Fire Engineering
**Glazing Consultant:** Billings Design Associates
**Special Consultant:** P. Bell

## Spandau Station, Berlin

**Structural Engineers:** Schlaich, Bergermann & Partner
**Architect:** von Gerkan Marg & Partner (Meinhard von Gerkan, Hubert Nienhoff)
**Client:** Deutsche Bahn AG

## Tower of Babel, Millennium Dome, Greenwich, London

**Engineers:** Atelier One (Neil Thomas)
**Architect:** Mark Fisher
**Contractors:** Sheetfabs (Nottingham) Ltd
**Client:** Millennium Dome

## Pavilion, Broadfield House Glass Museum, Kingswinford, West Midlands

**Structural Engineers:** Dewhurst McFarlane & Partners, London
**Chartered Architects/Designers:** Design Antenna (DA/da), Richmond (Partner: Brent Richards, Project Architect: Robert Dabell)
**Client:** Metropolitan Borough of Dudley, Broadfield House Museum
**Consulting Services Engineer:** Robert Scott Associates
**Quantity Surveyors:** The Dudley Design Partnership
**Main Contractor:** I.B Construction Ltd.
**Glass Manufacturers:** Saint Gobain/Solaglas Glass (UK), F.A Firmans Glass Installers

**Lighting:** Martini Lighting (UK), Optelma Lighting

**Flooring:** Stone Age

**Museum Displays:** Media Projects, McAndroids (with Design Antenna)

## Orangery, Prague Castle, Prague

**Structural Engineers:** Techniker Ltd (Matthew Wells)

**Architect:** Eva Jiricna Architects

**Client:** Prague Castle Management Office

**Mechanical Engineer:** Zemlicka P.K.

**Structural Checking Engineer:** Czech Technical University, Faculty of Civil Engineering

**Co-checking Engineer/Architect:** A.E.D s.r.o

**Project Manager:** Glockner Praha s.r.o

**Contractor:** Seele CZ s.r.o

## City of Science, Valencia

**Structural Engineer and Architect:** Santiago Calatrava (Prince Felipe Science Museum, Arts Palace, Hemispheric Imax Dome, Umbracle); Felix Candela (Oceanographic Park)

**Client:** Civis Project Management, Polytechnic University of Valencia

**Contractors:** FCC; ACS; NECSO; Sedesa; Dragados

## Waterland, Burgh-Haamstede

**Engineer:** Ingenieursbureau Zonneveld

**Client:** Waterland Neeltje Jans

**Sound Environment:** Edwin van der Heide, Victor Wentinck

**Sensors:** Bert Bongers

**Installations:** ABT

*Freshwater pavilion*:

**Architect:** NOXArchitekten (Lars Spuybroek)

**Light Spine:** euroGenie

**Projections:** Instituut Calibre

*Saltwater pavilion*:

**Architect:** Oosterhuis.nl

**Lighting:** Kas Oosterhuis, Arjen van der Schoot, Fairlight

**Virtual Worlds:** Kas Oosterhuis, Menno Gouwens, Eline Wieland, Károly Toth, Green Dino

## Guggenheim Museum, Bilbao

**Structural Engineers:** Skidmore Owings & Merrill, Chicago

**Architect:** Frank O. Gehry & Associates

**Project Principal:** Randy Jefferson

**Project Manager:** Vano Haritunians

**Project Architect:** Douglas Hanson

**Project Designer:** Edwin Chan

**Client:** Guggenheim Museum Foundation, Bilbao

**Mechanical Engineers:** Cosentini Associates, New York

**Lighting:** Lam Partners, Boston

**Acoustics & Audio Visual:** McKay, Connant, Brook Inc.

**Theatre:** Peter George Associates

**Security:** Roberti Bergamo E.A Italy

**Curtain Wall:** Peter Muller Inc., Houston

**Elevator:** Hesselberg Keese & Associates

**Contractors:** Cimentaciones Abando (Foundations), Ferrovial/Lauki/Urssa (Steel and Concrete Structure), Construcciones y Promociones Balzola (External Building), Ferrovial (Interiors, Building Systems, Site Works)

## Carlos Moseley Music Pavilion, New York

**Structural Engineer:** FTL Happold

**Client**: The Metropolitan Opera Association; The Philharmonic Symphony Society of New York, Inc.; Department of Cultural Affairs

**Theatre Design**: Peter Wexler Inc.

**Acoustics**: Jaffe & Holden

**Engineer**: McLaren Engineering Group

**Contractors**: Quickway Metal Fabricators, Inc.; The Fabric Shop; Bash Lighting

## London Eye, London

**Structural, Civil, Marine & Mechanical Engineers:** Atelier One, Beckett Rankine Partnership, Dewhurst McFarlane & Partners, Infragroep, Ove Arup & Partners (Jane Wernick), Sigma Plastique, Tony Gee & Partners

**Architects:** Marks Barfield

**Client:** The London Eye Company

**Construction Manager:** Mace

**Checking Engineer:** Allot & Lomax

**Capsule Design:** Nic Bailey

**Environmental Design:** Loren Butt

**Landscape Architect:** Edward Hutchison

**Leisure Consultant:** Ray King

**Lighting Artist:** Yann Kersale AIK

**Subcontractors:** Hollandia, Mercon, Skoda Steel, Mannesman, Tensotecci (Main Steelwork), Corius (Steel Rim), Poma, Sigma, Semer (Capsules), Sunglass Glaverbel (Capsule glazing), Tilbury Douglas Construction (Civil Engineering), T. Clarke, DAL (Services), Littlehampton Welding (Boarding Platform), LB Securities (Fire Shutters), Latchways (Maintenance Equipment), Waterers (Landscape)

## Stock Exchange and Chamber of Commerce (Ludwig Erhard Haus), Berlin

**Structural Engineers:** Whitby & Bird, Specht Kalleja & Partner GmbH

**Architect:** Nicholas Grimshaw & Partners

**Client:** Berlin Chamber of Commerce

**Quantity Surveyor:** Davis Langdon & Everest, Mott Green Wall

**Contractor Team/Services Engineers:** RP + K Sozietät

## Educatorium, Utrecht University, Utrecht

**Engineer:** ABT Adviesbureau voor Bouwtechniek BV, Robert Nijsse, Frans van Herwijnen, Velp-Delft

**Architect:** Office for Metropolitan Architecture (OMA), Rem Koolhaas, Christophe Cornubert

**Client:** Universiteit Utrecht, Marianne Gloudi

Services Engineer: Ingenieursburo Linssen, Henk Knipscheer, Amsterdam

**Acoustics:** TNO-TUE

**Facades:** Robert-Jan van Santen

**Installations:** FBU

**Projectioncel:** Curve

**Artists:** Joep van Lieshout, Andreas Gursky

**Ecological Consultant:** W/E Advisers duurzam bouwen

**Construction:** BAM Bredero, Bunnick/A. de Jong airconditioning bv, Schiedam/Ergon Electric bv, Utrecht/GTI Rotterdam-Capelle bv, Rotterdam /Lichtindustrie Wolter & Drost-Evli bv, Veenendaal.

## Traversina and Suransuns footbridges, Viamala Gorge

**Structural Engineer:** Conzett Bronzini Gartmann AG

**Client:** Verein KulturRaum Viamala

**Main Contractor:** Flutsch Holzbau

**Steel Cable:** Riss AG

**Steel Joints (Stahlknoten):** Romei AG

**Concrete Contractor:** V. Luzi

## Solférino Bridge, Paris

**Engineer & Architect:** Marc Mimram

**Client:** Ministry of Culture (Ministère de la Culture); Ministry of Public Works, Housing and Transport (Ministère de l'Equipement, du Logement et des Transports) in collaboration with the Louvre (Etablissement Public du Grand Louvre) and the Public Foundation for Cultural Works (Nouvel Etablissement Public de Maîtrise d'Ouvrage des Travaux Culturels)

**Consultant:** Sogelerg

## Expo Roof, Hanover

**Structural Engineers:** IEZ Natterer GmbH

**Architect:** Herzog & Partner BDA

**Client:** Deutsche Messe AG

**Project Management:** Assman Beraten und Planen GmbH, Hamburg

**Vibration Report:** Technische Universität München, Institut für Tragwerksbau

**Proof Engineers for Structural Analysis:** Ingenieursbüro Speich-Hinkes-Lindemann

**Colour Design:** Prof. Rainer Wittenborn

**Lighting Design:** Ulrike Brandi Licht

**Membrane planning and Engineering:** IF Jorg Tritthart, Dr -ing. Harmut Ayrle, Reichnau/Konstanz

**Soil Report:** Dr -ing. Melhorst & Partner

**Fire Protection:** Hosser, Hass & Partner

**Foundations:** Renk Hortsmann Renk

**Surveying Services:** Drecoll v Berckefeldt

**External Works:** Dieter Kienast, Vogt Partner

## Hall F, Charles de Gaulle International Airport, Terminal 2, Paris

**Structural Engineer:** Paul Muller

**Architects:** Paul Andreu, Jean-Michel Fourcade in association with Anne Brison, Valérie Chavanne

**Project Managers:** Patrick Trannoy, Michel Jean-François, Patrice Hardel, Nathalie Roseau

**Design Office:** Aéroports de Paris, Architecture and Engineering Department

**Steelwork and Facades:** RFR

**Air-conditioning:** Cabinet Trouvin

**Project Co-ordination:** Roland Micard, Philippe Gourcerol, Norbert Marduel

**Main contractors:** Bouygues (Roads), Spie-Batignolles (structural work) Viry-Fischer (glazing), Harmon CFEM (landside walls and gable), Portal (metal millwork), Chambon (masonry), Santerne (high-voltage electricity), I.C.E (air conditioning), G.U.L (roofing), Dezellus M.I. (architectural finishings), France Sols (marble floors), Saint Gobain Vitrages (glass), Usinor (steel), Ugine (stainless steel), Vieille Montagne (zinc)

# Index

# Picture credits & Author's acknowledgements

Architectural Association Photo Library/Ulrich Kerber (18); Hans Bach (107 bottom left, 108); Richard Barnes/ESTO (81); Bau Bild berlin (109 bottom left); Richard Bryant/ARCAID (12, 125–131); Courtesy Burnstein Associates (79–80); Jeremy Cockayne/ARCAID (13); Peter Cook/VIEW (21, 52–61); Richard Davies (98–102 bottom, 103); Michel Denance/Archipress (211, 215); Denis Finnin (77); Chris Gascoigne/VIEW (24 top left); Berengo Gardin Gianni (95); Dennis Gilbert/VIEW (119); Jeff Goldberg/ESTO (156–159); Hiroyuki Hirai (44–50); Peter Horn (89–93); Nick Huften/VIEW (63–64); Keith Hunter (72–75 bottom); Courtesy IDOM (151 top, bottom right); Mark Jones /ARCAID (19); Nicholas Kane/ARCAID (97); Ian Lambot (163, 168–170); Pawel Libera/ARCAID (24 bottom); Katsuhisa Kida (41 top left, 83–87, 120–124); Jay Langlois (20); John Edward Linden (23); Peter Mackinven/VIEW (66–67, 69–71, 73, 75 top, 102 top); Duccio Malagamba (148–151 bottom left, 152–155); Yoshiharu Matsumura (30–37); Paul Maurer (212); Moises Puente, Schenk & Campbell (24 top right, 193–195); J. M. Monthiers (198, 202); Punctum/H.-Chr. Schink (6, 105, 106, 107 bottom right, 109 top); Paul Raftery/VIEW (11, 197, 201); Mandy Reynolds (22); Ralph Richter/architekturphoto (205–207); Diana Scrimgeour (110–117); Grant Smith (39, 41 top right, 42–43); Tiefbauamt Gaubunden, Switzerland (15); Jeanette Tuschudy (187–191); Morley von Sternberg/ARCAID (10); drawings by Matthew Wells (192, 193 top, 201 bottom); Hans Werlemann (180–185); Courtesy Wilkinson Eyre (25, 26–27); Jens Willebrand (173–178); Alan Williams/ARCAID (17); Nick Wood (165, 167); Javier Yaya/CACSA (134–139)

I have to particularly thank Matthew Wells, founder of Techniker, whose insight was crucial. My understanding of the engineer's task has been enriched over the years by two extraordinary people, Frank Newby, former chief of Samuely and Partners and Ted Happold, who devoted the second half of his life to educating engineers among architects and other members of the building team. They are sorely missed. There are many others including another great structural engineer, Tony Hunt; Ian Liddle, an immensely creative disciple and partner of Ted Happold; the wise Derek Sugden of Arups; Tim McFarlane who has proved Fermat's Last Theorem for all the even numbers; Henry Bardsley, disciple and partner of the late great Peter Rice; Mark Whitby and all the engineers who appear in this book.

Books are a collaborative thing and I have to thank my terrific Laurence King editor Liz Faber, ace research co-ordinator Jennifer Hudson and longsuffering editorial director Philip Cooper. Project editor Lucy Trench and I happily discussed syntax and meaning by e-mail and book designer Gavin Ambrose got things excellently right more or less first time. My marvellous agent, Shelley Power, remained supportive throughout and dear Rosey cheerfully put up with it all. Again.